海绵城市
——景观设计中的雨洪管理

Sponge City
— STORMWATER MANAGEMENT IN LANDSCAPE DESIGN

戴滢滢 编

江苏凤凰科学技术出版社

图书在版编目（CIP）数据

海绵城市：景观设计中的雨洪管理 / 戴滢滢编. --
南京：江苏凤凰科学技术出版社，2016.3
　ISBN 978-7-5537-6170-1

Ⅰ. ①海… Ⅱ. ①戴… Ⅲ. ①城市景观－景观设计－
研究 Ⅳ. ①TU-856

中国版本图书馆CIP数据核字(2016)第035096号

海绵城市——景观设计中的雨洪管理

编　　　者	戴滢滢
项 目 策 划	孙　爽　曹　蕾
责 任 编 辑	刘屹立
特 约 编 辑	孙　爽

出 版 发 行	凤凰出版传媒股份有限公司
	江苏凤凰科学技术出版社
出版社地址	南京市湖南路1号A楼，邮编：210009
出版社网址	http://www.pspress.cn
总　经　销	天津凤凰空间文化传媒有限公司
总经销网址	http://www.ifengspace.cn
经　　　销	全国新华书店
印　　　刷	利丰雅高印刷（深圳）有限公司

开　　　本	965 mm×1 270 mm 1/16
印　　　张	19.5
字　　　数	250 000
版　　　次	2016年3月第1版
印　　　次	2016年3月第1次印刷

标 准 书 号	ISBN 978-7-5537-6170-1
定　　　价	320.00元（精）

图书如有印装质量问题，可随时向销售部调换（电话：022-87893668）。

海绵的哲学

"海绵城市"既是一种城市形态的生动描述,也是一种雨洪管理和治水的哲学观和方法论。

"海绵城市"是指建立在生态基础设施之上的生态型城市。生态基础设施有别于传统以单一目标为导向的"灰色"工程性基础设施;其以综合生态系统服务为导向,运用生态学的原理和景观设计的方法以及渗、蓄、净、用、排等关键技术,实现以城市内涝和雨洪管理为主且同时包括生态防洪、水质净化、地下水补给、棕地修复、生物栖息地保护和恢复、公园绿地建设及城市微气候调节等综合目标。

"海绵城市"符合中国独特的地理气候特征,以中国悠久的水文化遗产为基础,并融合了当代国际先进的雨洪管理技术和生态城市思想,是一套完整的方法论和技术系统。

以"海绵"比喻一个富有弹性且以自然积存、自然渗透、自然净化为特色的生态型城市,是对工业化时代机械的城市建设理念及水资源、水系统的错误认识的反思,蕴含深刻的哲理,是对简单工程思维的"反叛"。这种"反叛"的哲学集中体现在以下几个方面。(注:以下部分内容首次发表于《景观设计学》期刊2015年14(2)期,题为《海绵的哲学》,收入此处,权当本书之序)

第一,完全的生态系统价值观,相对于功利主义的片面的价值观。稍加观察不难发现,人们对待雨水的态度实际上是非常功利的,而且非常自私。砖瓦场的窑工,天天祈祷第二天是个大晴天;久旱之后的农人,天天到龙王庙里烧香,祈求天降甘霖;城里人把农夫的甘霖当祸害。同类之间尚且如此,对于诸如青蛙之类的其他物种,就更无关怀和体谅可言了。"海绵"的哲学是包容,对这种以人类个体利益为中心的雨水价值观提出了挑战,宣告:天赐雨水都是有其价值的,因为它不是对某种人、某个物种有价值,而是对整个生态系统有天然的价值。人作为这个系统的有机组成,是整个生态系统的必然产物和天然受惠者。所以,每一滴雨水都有它的含义和价值,"海绵"珍惜并留下每一滴雨水。

第二,就地解决水问题,而不是将其转嫁给异地。把灾害转嫁给异地,是一切现代水利工程的起点和终点:诸如防洪大堤和异地调水,都是把洪水排到下游或对岸,或把干旱和水短缺的祸害转嫁给无辜的弱势地区和群体。"海绵"的哲学是就地调节水旱,而不转嫁异地。它启示我们运用"适应"的智慧,就地化解矛盾。中国古代的生存智慧是将水作为财,就地蓄留,无论是来之屋顶的雨水,还是来之山坡的径流,因此有了农家天井的蓄水缸和遍中华大地上的陂塘系统。"海绵"景观既是古代先民"适应"旱涝的智慧,也是地缘社会和邻里关系和谐共生的体现,更是几千年以生命为代价换来的经验和智慧在大地上的烙印。我家的多位祖先就曾试图将白沙溪上游的一道水堰提高寸许,以便能灌溉更多田亩,而与邻村发生械斗,最终不幸丧命。因此,在自家的田里挖塘蓄水,调节旱涝,就地解决水旱之灾,是最好的途径。

第三,分散式的民间工程,而非集中式的集权工程。中国的常规水利工程是集国家或集体意志办大事的体现。从大禹治水到长江大坝,无不体现国家意志之上的工程观。这也是中国数千年集权社会制度产生和发展的重要原因之一。在有的情况下这是有必要的,如都江堰水利工程,得益于其对自然水过程因势利导的哲学和工程智慧,而沿用至今,福泽整个川西平原。然而,集中式大工程,如大坝蓄水、跨流域调水、大江大河的防洪大堤、城市的集中排涝管道等失败案例非常之多,典型的工程如三门峡水库、阿斯旺水库等。从当代的生态价值观来看,与自然水过程相对抗的集中式工程并不明智,也往往不可持续。而民间分散式、民主式的水利工程往往具有更好的可持续性。在广袤的中华大地上,古老的民间微型水利工程,诸如陂塘和水堰,至今仍然充满活力,受到乡民的悉心呵护。然而,令人非常遗憾的是,这些千百年来滋育中国农业文明的民间水利遗产,在当代却遭到强势的国家水利工程的摧毁。"海绵"的哲学是分散,由千万个细小的单元细胞构成一个完整的功能体,将外部力量分解吸纳、有机消化。因此,我们呼吁珍惜并呵护民间水工遗产,提倡分散式、民主式的微型水利工程。这些民间水工设施不对流域的自然水过程和水格局造成破坏,且构筑了能满足人类生存与发展需求的伟大的国土生态海绵系统。

第四,慢下来而非快起来,滞蓄相对于排泄。将洪水、雨水快速排掉,是当代排洪排涝工程的基本哲学。所以,三面光的河道截面被认为是最高效的,裁弯取直被认为是最科学的,河床上的树木和灌草必须清除以减少水流阻力,被认为是天经地义的。这种以"快"为标准的水利工程罔顾自然水过程的系统性和水生态系统的主导因子的完全价值,以致加强、加速了洪水的破坏力,将上游的灾害转嫁给下游,将水与其他生物分离,将水与土地分离,将地表水与地下水分离,将水与人和城市分离,导致地下水得不到补充，土地得不到滋润，生物栖息地不断消失。"海绵"的哲学是将水流慢下来，让它变得心平气和而不再狂野恐怖，让它有机会下渗和滋育生命万物，让它有时间净化自身，更让它有机会服务于人类。

　　第五，弹性应对，而非刚性对抗。当代治水工程忘掉了中国古典哲学的精髓——以柔克刚，却崇尚"严防死守"的对抗的哲学，遍中华大地上已没有一条河流不被刚性的防洪堤坝所捆绑，原本蜿蜒柔和的水流形态如今都变成直泄刚硬的排水渠。千百年来的防洪抗洪经验告诉我们，当人类用坚固的防线将洪水逼到墙角之时，洪水的破堤反击便指日可待，此时的洪水便成为摧毁一切的猛兽，势不可挡。海绵的哲学是弹性地应对外部冲力，化对抗为和谐共生，正所谓"退一步海阔天空"。如果我们崇尚"智者乐水"的哲学，那么，水的最高智慧便是以柔克刚。

　　"海绵"的哲学强调将有化为无，将大化为小，将排他化为包容，将集中化为分散，将快化为慢，将刚硬化为柔和。在"海绵城市"成为当今城市建设一大口号的今天，只有深刻理解其背后的哲学，才能使之不会沦为某些城市、工程建设公司及规划设计师的新的形象工程和牟利的幌子，从而避免新一轮水生态系统的破坏。老子所谓"道恒无为，而无不为 (Nature does not hurry, yet everything is accomplished)"正是"海绵"哲学的精髓。

北京大学建筑与景观设计学院

2015 年 11 月

THE PHILOSOPHY OF THE SPONGE

Sponge City is a vivid description of an urban form, and the philosophy, and methodology of stormwater management and flood control practice.

Sponge City is a kind of eco-city based on ecological infrastructure. The eco-infrastructure is different from traditional gray infrastructure of which is single targeted and mechanically constructed. It's an integrated service-oriented ecosystem, by employing ecological principles, landscape architecture approaches, and the key techniques of infiltration, storage, purification, utilization and discharge, to accomplish the main goal of urban waterlogging prevention and stormwater management, and to achieve the comprehensive targets such as ecological flood control, water purification, groundwater recharge, brownfield restoration, habitat protection and restoration, green parks construction and urban micro-climate regulation, etc.

Sponge City is a theoretical, methodological and technical system integrated with contemporary internationally advanced stormwater management technologies and ecological urbanism thoughts, which was put forward in the context of China's unique geographical and climatic characteristics and on the foundation of China's historic cultural heritage.

Sponge is a metaphor, refer to the elastic eco-city accumulate and infiltrate runoff naturally. It's a prudent reflection on the mechanical city construction idea in the industrial age and the errors made in water resource utilization and hydraulic system protection. This idea is critical to the simple engineering concepts, and contains sophisticated philosophy. This rebellious thoughts embodied in the following aspects (Note: The following section was first published in *Landscape Architecture*, 2015, 14 (2): 4-9, named as *Philosophy of Sponge*, and used as the preface in this book.

First, complete eco-systematic value, rather than partial utilitarian value. It only takes a little observation to find the human's selfish and pragmatic attitude towards rainwater. The workers in the brick kiln prayed for the sunny day all year long. The peasant suffered from draughts went to the Dragon King Temple to burn incense to pray for rains day after day. And the citizens treated the rain as a nuisance while the farmers regarded it as the source of life. Such a huge gap existed among human beings themselves, not to mention the alien attitudes towards rain in different species. Taking the frogs for example, the concerning and caring is wretchedly inadequate. The philosophy of sponge is focus on containment. The idea is a great challenge toward human beings' self-engrossed value in rainwater. To convey sponge philosophy is tantamount to proclaim: the rain is bestowed by heaven with its inherent value. It is intrinsically valuable to the whole eco-system, not just for some specific people, belief or species. As an organic part of the system, human beings are the inescapable productions of the eco-system and the best beneficiary of the nature. So, each drop of rain has its meaning, has its value. Sponge treasures and conserves every drop of rain.

Second, try to solve water problems on-site, rather than transferring the issue to other area. Every modern water conservancy project originated from and ended in the intention of delivering the disasters to the other places. For instance, the methods like flood control dyke and the long distance water transfer, all discharged the floods downstream or to the opposite bank, or impose the damage from draughts & water shortage to the innocent region and vulnerable communities. The philosophy of the sponge is to regulate humidness and dryness locally without shifting it. It sparked us on the brightness of adaptation, and inspired us to solve the problem on-site. In ancient Chinese survival wisdom, water is a wealthy and fiscal sign. No matter the rain from the roofs, or the runoff down the hills, they are all collected in places. This is the most convictive explanation why the water harvest tank in the farmers' household courtyard so ubiquitous and why the pond-trench systems on the majority of land in China so popular. This Sponge Landscape is our ancestors' ancient wisdom adapted to drought and flood, but also a harmonious co-existent refection of the relations in geographical society and geopolitical neighborhood. It's the experience and wisdom imprinted on earth at a price of millions of generations' sacrifice for thousands of years. Many elder members in my family lost their lives in the fight with the neighboring village. This knife fight only intrigued by the intention to lift the dam several inches up on the upstream of White Sand Stream, in order to expand the irrigation area for only a little longer miles. Therefore, the best solution to deal with the wet &dry disaster is to dig the storage pond on one' own land, and regulate the draught and flood just on-site.

Third, decentralized civil construction, rather than centralized authoritarian project. China's water conservancy projects are sets of conventional programs reflect the will of the county and community. This kind of construction works range from the water management by Dayu in ancient era to the Yangtze River dam in modern time. And this is one of the direct reasons of the origin and development of central authoritarian social system in China for thousand-years-history. In some circumstances it is necessary, such as Dujiangyan Irrigation Project, due to the conceptual and engineering wisdom of guiding the water in the light of its natural trend, the project still functioning well till today, and benefit the entire Sichuan flatland. But the failure cases in the centralized large projects are plentiful, such as, water storage dams, cross-regional water transfer projects, flood prevention weir of major rivers, centralized drainage pipe system in cities. The most typical projects are the Sanmenxia Reservoir, Aswan reservoirs, etc. From the view of contemporary ecological values, the centralized projects confronted with the natural process are both inadvisable and unsustainable. To the contrary, the decentralized or democratic folk hydraulic engineering work usually holds a better sustainability. The ancient miniature water conservancy projects scattered on vast land of China conserved intact under the good care from the villagers. Such as ponds and weirs, still full of vitality till nowadays. Very unfortunately, these historic folk hydraulic heritage once have bred the Chinese agricultural civilization was destroyed by national water conservancy projects forcibly. The philosophy of sponge is decentralization. A functional entity is made up by millions of tiny cells. It can permeate and absorb the external strength in spontaneous way. Therefore, we appealed for cherish and care to the civilian water conservancy heritage, and advocate domestic and dispersed micro-irrigation project. These decentralized civilian water management facilities contributes to the great country's ecological sponge system without any harm to the natural watershed process and pattern, but meet human beings survival and developing needs.

Fourth, slow down, rather than speed up, detain &retain, rather than discharge &blow-off. Drain the storm water as fast as possible. It's regarded as the basic philosophy of contemporary flood drainage project. So the three-facet polished watercourse is the most efficient. The straightened river way is the most scientific. Even the trees, shrub and grass on the riverbed, are bound to be removed in order to reduce the flow resistance. This fast-speed oriented guidance turns a blind eye on the systematic feature and dominant factors' value of the water pattern and hydraulic network. As a result, the destructive power of floods strengthened and devastated, the hazard shifted from upstream to downstream. Water is separated from earth, other creatures, human beings and cities. The linkage between surface water and groundwater was decreased and cut off, the underground water can't be recharged, the earth can't be moistened, and the biological habitat gradually disappeared. The philosophy of sponge is to slow the water flow down, calm it, ease it, let it no longer wild and horrifying, as well as to provide the water with the opportunity to permeate in land, nourish the life, purify itself, have more chance to serve human.

Fifth, hold on flexibly, rather than resist rigidly. Contemporary engineering flood management thoughts overlooked the classical Chinese philosophical essence--overcoming firmness by gentleness, but stick on the concept of resistance—defending resolutely. The inflexible weirs and rigid dams trussed up the rivers in China. None of them could escape the misfortunate destiny. The originally soft and meandering flow declined to rigidly straight-rowed drainage trench. Thousand-year experiences of flood-control and flood-prevent experiences tell us, if human cornered the flood by strong defense line forcefully, the floods would become fierce monsters with tremendous and unstoppable destructive power. The counterattack from spate would be impending, while the preventing dams would inevitably collapsed and destroyed. Sponge copes with external momentum through the philosophy of flexibility and Changes the resistance and confrontation into harmonious coexistence. As the saying goes, taking a step back as boundless as the sea and sky. If we advocate the philosophy of water delights the wise man, then, highest wisdom of the water might be conquering the unyielding with the yielding.

The philosophy of sponge emphasis on turning something into nothing, turning huge to small, turning the exclusion to tolerance, turning centralization to decentralization, turning fast to slow, turning hardship into softness. Sponge City becomes a major slogan for urban construction nowadays. If not to perceive and study the philosophy profoundly, the Sponge City construction could be reduced into a signboard of the vanity projects and profit-making projects by some cities, engineering companys as well as planners &designers. Besides, this practice also could avoid a new round damage towards the ecosystems of water. The founder of Taoism (Lao Tzu) once said prudently, Nature does not hurry, yet everything is accomplished, which has illustrated the essence in the philosophy of sponge.

Peking University, the School of Architecture and Landscape Design
Kongjian Yu
Nov. 2015
(Translated by Zhangxu)

序 海绵城市 —— 让城市回归自然

城市化在给城市道路、交通、住房等基础设施带来较大改善的同时，也引起了气候、水文、植被以及环境质量的改变。伴随着城乡双重城镇化的无序蔓延，土壤污染、水污染、自然人文生态系统崩解，这使人们不禁无限感慨，"融不进的城市，回不去的故乡"。

顾名思义，"海绵城市"是指城市像海绵一样，在适应环境变化和应对自然灾害等方面具有良好的"弹性"，下雨时吸水、蓄水、渗水、净水，有需要时将蓄存的水"释放"并加以利用。其本质是解决城镇化与资源环境的协调和谐，目标是让城市"弹性适应"环境变化与自然灾害、转变传统的排水防涝思路，运用低影响开发理念，充分发挥河流、山体、城市绿地、道路、建筑对雨水的吸纳、蓄渗和缓释作用，利用透水铺装、植被渗沟、下凹绿地和湿地水体等景观元素，重建接近自然的水循环过程，使开发前后的水文特征基本不变，实现水资源的自然积存、自然渗透、自然净化。

在国外先进的理念中"海绵城市"理念已经从传统简单的管渠排水工程技术的层面发展到与景观生态设计紧密结合的雨洪生态管理的层面，强调采取生态和近自然的生态措施，充分发挥城市自然生态系统在涵养水源、调蓄雨洪、净化径流污染、水质保护、雨水资源化利用等方面的生态系统综合服务价值，通过科学合理的规划设计，协调自然和人工景观，维护和提升城市自然水文循环过程，进而实现城市的可持续发展。如美国、英国、澳大利亚等发达国家已经形成了相对完善且符合本国技术法规要求的"海绵城市"管理模式体系，并将其很好地结合与应用于城市景观和基础设施的规划设计与建设中。我国真正意义上的城市雨洪管理开始于20世纪80年代，发展于90年代，目前呈现出良好的发展势头。其核心是截污、减排、滞留、利用、生态。

"海绵城市"主要由自然系统、半自然系统、人工系统三部分构成。自然系统主要包括河流、湖泊、湿地、山地、森林等；半自然系统主要包括水库、坑塘、郊野公园、农田、城市绿带、防护林带等；人工系统主要包括建筑雨水、绿色屋顶、透水性地面等。构建"海绵城市"，让城市回归自然，可通过以下三个途径。

(1) 充分保护区域原有生态系统，并对已经受到破坏的自然环境及水体进行修复。

首先，从"海绵城市"的构成要要素着手，寻找对地表径流量产生重大影响的自然、半自然斑块，对水文影响最大的斑块需要严加保护。其次，构建生态廊道。生态廊道使分散的、破碎的斑块有机地联系在一起，成为更具规模和多元化的生物栖息地和水生态水资源涵养区，为生物迁移、水资源调节提供必要的通道与网络。第三，划定全区生态控制线，明确城市开发建设的边界，防止城乡建设用地无序蔓延，引导城市健康发展及精明增长。最后，对已经受到破坏的自然环境及水体进行生态修复，采用各种技术手段，提高水生态系统的自然修复能力。

(2) 开展并加强城市规划区内海绵城市规划设计与改造。

"海绵城市"需要城市各层级、各相关专业规划发挥控制和引领的总用。

城市的总体规划阶段应注重理念与方法的创新，以低影响开发雨水系统为重要手段，结合城市生态保护、土地利用、水系、绿地系统、市政基础设施、环境保护等相关内容，因地制宜地确定城市年径流总量控制率和对应的设计降雨量目标，制订城市低影响开发雨水系统的实施策略，并确定重点实施区域。

城市专项规划，包括城市水系统、绿地系统、道路交通等基础设施专项规划。其中，城市水系统规划涉及到供水、节水、污水（再生利用）、排水（防涝）、蓝线等要素；城市绿地系统规划应在满足绿地生态、景观、游憩等基本功能的前提下，合理地预留空间，并为丰富生物种类创造条件，对绿地自身及周边硬化区域的雨水径流进行渗透、调蓄、净化，并与城市雨水管渠系统、超标雨水径流排放系统相衔接。道路交通专项规划，要协调道路红线内外用地空间布局与竖向，利用不同等级道路的绿化带、车行道、人行道和停车场建设雨水滞留渗设施，实现道路低影响开发控制目标。

城市的控制性详细规划阶段着重落实城市总体规划及相关专项规划，确定低影响开发控制目标与指标，因地制宜，落实涉及雨水渗、滞、蓄、净、用、排等用途的低影响开发设施用地；并结合用地功能和布局，分解和明确各地块单位面积控制容积、下沉式绿地率及其下沉深度、透水铺装率、绿色屋顶率等低影响开发主要控制指标，指导下层级规划设计或地块出让与开发。

具体来讲，要结合城市水系、道路、居住区、绿地与广场等空间载体，建设低影响开发的雨水控制与利用系统。

城市水系是最直接且最理想的雨水调蓄、排放的场所。在规划与改造时，应在充分保留原有水系的同时，尽量结合实际情况，开挖沟渠，连通各个水系，形成完整的城市水系。水岸设计除必须考虑防洪安全等问题的硬质铺装外，还应多设置生态驳岸，提高水系吸收、净化雨水的能力。

城市道路是雨水等地面径流汇聚的主要场所，道路两侧人行道可采用透水式铺装，道路两侧绿化带及行车中分带可适当下沉，同时还可将雨水通过向周边公园绿地等自然环境引流排放，以更好地储存和吸收雨水。

城市绿地与广场是人们休闲和娱乐的主要目的地，也是雨水主要汇集的场所。绿地应多种植被相结合，在营造良好的自然景观的同时，通过地被等灌木植物根系更充分吸收雨水。广场铺装应尽量采用透水铺装，充分考虑引导雨水汇集、吸收的作用，完善城市水循环系统。

居住区其内部道路、人行道、休闲广场应采用透水铺装。在规划建设绿地时，应尽量将绿地采用下沉式设计，低于地平面的设计可更有效地储存、吸收、滞留雨水，并可作为小区内部的水系景观。对于建筑，则应多采用屋顶绿化的方式，在吸收、滞留雨水的同时，也可起到美化城市景观、缓解城市热岛效应的作用。

(3) 建立"海绵城市"管理保障实施机制。

建立"海绵城市"管理保障实施机制，可通过制定各种法律法规、加强机制改革、强化管理监督等的措施来促使"海绵城市"逐步规范化、法制化。首先，应制定并完善"海绵城市"建设的法律法规及政策体系，为"海绵城市"建设提供制度保障。其次，各地方政府应改革当前的城市雨水管理机制体制，包括建立与考核相关的奖惩机制、"海绵城市"建设的投融资机制等，在体制改革上，应改变当前城市规划设计不透明、缺乏社会监督的管理体制，在"海绵城市"建设中，从项目立项、规划设计到建设施工的整个过程应强化社会公众的监督。

"海绵城市"让城市回归自然。在这个时代，自然就是我们。未来是我们选择的结果，而不是我们必然的命运。试想一下，到2200年或2500年时，我们的子孙后代可能会把我们比作把地球仅仅用来中途停靠加油的外星人，或者甚至将我们定义为摧毁自己家园的野蛮人。生活在这个时代意味着要营造一方能与地球的生物资源共同成长而不是无限掠夺的文化净土。

"海绵城市"作为一种新型的城市建设模式，其理论尚处于探索和发展阶段。我们应该从更深远的层次理解其概念和内涵，找到符合我国不同地域特色的具体实施策略并制定一系列规范标准，促进其理论体系的完善和成熟，促使城市回归自然。我们的工作任重而道远。

2015 年 9 月

SPONGE CITY—BRING THE CITY BACK TO THE NATURE

The process of urbanization has observed considerable changes in city roads, transportation, housing and other infrastructures, and the deteriorated quality of the climate, hydrology, vegetation, and environment. Soil pollution, water contamination, the paralysis of the natural ecological network, the collapse of the cultural ecological system, these issues happened frequently, accompanied with the disordered sprawl in both urban & rural areas. People cannot help but reflect, what a city without social ladders, what a hometown without retrograde steps.

Just as its name implies, the feature of Sponge City should resembled the feature of sponges, which require them to embody enough "flexibility" and "reliance" in face of the climate change and natural hazard, etc. The city can absorb, harvest, filter and purify water when it rains, while "release" the stored rainwater when in need. Its essence lies in solving the disharmony between urbanization and environmental resources. Its goal is to make cities adapted to environmental changes in a flexible manner and confront with natural disasters in a resilient way,to transform the traditional stereotypes on city drainage and waterlogging prevention, and implement low-impact urban development practice,to give a full play on rivers, mountains, urban green spaces, roads, buildings in terms of the function of rainwater absorption, filtration and detain & retain. to reconstruct the natural-like water circulation by employing the landscape elements, such as permeable paving, filtration bio-swale, sunken green space, wetlands and other water bodies,as well as to stabilize the hydraulic characteristics in every urban developing phase, and accumulate, infiltrate and purify the water resource in a natural way.

In the advanced foreign concept, the philosophy of Sponge City initiated by simple and traditional pipe & tube drainage engineering techniques, which are substituted by the ecological landscape design integrated with the idea of ecological stormwater management. Highlighting in employing ecological and biomimetic measures, making full use of the comprehensive service value lies in the natural eco-system, which could carry on the processes like water storage, rainwater regulation, runoff pollutants purification, water quality protection, etc. enhancing natural hydrological cycle in city and maintaining the long-lasting sustainable city development by scientific plan and rational design, and balancing the natural a and artificial landscape. Several developed countries have already formed the comparatively consummate Sponge City management system model, which appropriately suits the in-house technical law & regulation system. Besides, the model was perfectly combined with the plan of urban design and practiced through the design of infrastructure construction. The genuine Sponge City was emerged around the 1980s in China, developed and thrived in the 1990s, and maintained a growing momentum nowadays. The core of the Sponge City focus on reducing pollution, cutting exhausting, retention & detention, recycle & reuse, ecology, etc.

We believe that Sponge City is mainly composed of three parts, namely, natural systems, semi-natural systems and artificial systems. Natural systems mainly include rivers, lakes, wetlands, mountains, forests, and the things alike. Semi-natural systems consist of reservoirs, ponds, country parks, farmlands, urban green belts, shelter forest belts, etc. Artificial system is comprised of rainwater utilization facilities, green roofs, permeable pavement and so on. How to build a Sponge City? How to bring the city back to nature? You may find the answer in the following three main approaches.

(1)To render all-around protection to the original eco-system on target area, repair &restore damaged natural environment and water bodies.

First, to proceeds the work on the components of Sponge citiy. Finding out the natural &semi-natural patches, which have significant impact on the surface runoff. Strictly protecting the patches of which could exert great influence on the regional hydrology. Second, to build the ecological corridor. Ecological corridors linked the scattered & fragmented patches together organically. Apart from this, they also formed some scaled and diversified bio-habitats and water conservation areas, which provide the migration route for animals and aquatic network for water regulation. Third, to delineate the regional ecological control lines, and make a clear boundary on urban development, prevent the construction land sprawl in urban & rural areas, and spark the city with sound development and smart growth. Finally, to repair the damaged natural environment and restore the spoiled water bodies ecologically through various technical means, and improve natural water ecosystems' self-healing ability.

(2) Sponge City design & reconstruction in urban planning area.

"Sponge City" demands the control and guide from all levels of cities and all the disciplines of relevant professions.

The phase of comprehensive urban planning should focus on innovative ideas and methods, and on the stormwater management measures which emphasized on low-impact city development accompanied with the related contents of urban ecological protection, land use, hydrographic network, green systems, municipal infrastructure, environmental protection, formulate & regulate the control rate of annual total runoff volume and the corresponsive goal of the annual designed rainfall volume based on the local condition, set out the feasible & practical policies for the low-impact stormwater management system, and select the key implementation area.

Urban subject plan, include the specialized plan in the aspects of water system, green space system, road & transportation systems, and other infrastructure systems in the city. Among them, urban water system planning is intertwined with water supply, water conservation, wasting water (recycle & reuse), drainage (waterlogging prevention), the blue control line and other factors. The green space system planning should meet the basic functional requirements from the precondition of greenery ecology, landscape, recreation and other. Preserve the land and space rationally , as well as provide suitable condition for various species to bread & generate. Infiltrate, regulate and purify the runoff on site & from the surrounding hardened area, and integrate in urban stormwater pipe & channel drainage system and excessive runoff discharge system. As for the specialized plan of road & transportation system, it should not only aim in the coordination between the space layout inside and outside the red control line, but the management of the vertical space layers. We should make full use of the green belt, traffic pass, pedestrian way and the parking lot in different levels of roads, and build stormwater retention & detention facilities to achieve the low-impact development goal.

The phase of regulatory detailed urban plan focues on the implementation of the comprehensive plan and the specialized plan. Set off the controlled goal and standard for the low-impact development. Arrange the lands operated as the stormwater infiltration, detention & retention, purification, utilization and discharge space on the basis of the local condition, and in line with the land use and area layout. Decompose and definitude the low-impact criteria in every modular area, such as the plot ratio, green rate and the depth of the sunken green space, permeable paving rate and green roof rate, etc. all the subordinate plan and land transfer & development should follow these guidelines.

Specifically, to build the low-impact stormwater control & utilization system conjuncted with space carrier such as urban water system, roads, residential areas, green spaces and squares.

Urban water system is the most direct and ideal place to regulate and drainage stormwater. While sufficiently conserving the original water system, taking the site's situation into fully accounts, excavating the ditches and tunnels, connecting all the hydrographic net and forming a complete urban water system. The coast lines design must consider the hard pavement to prevent flood, and give a second thought on the ecological revetment, and improve the water system' capacity in absorbing and purifying storm water.

Urban road is the primary space for rainwater runoff gathering. The pedestrian ways on two sides of the roads could adopt the permeable paving style. The green belt on both sides and the central separation barriers could be properly carved with submergence to convey the stormwater runoff and discharged it into the surrounding green park and other natural environment, for the refined function of stormwater harvesting and infiltration.

Urban green space and plaza is the main recreational place for people, and also the main gathering place for rainwater. Green space design should pick a variety of vegetation, while create a favorable natural landscape but also absorb rainwater through the roots of ground cover plants and shrubs. The pavement on the square should use pervious ones to the greatest extent and fully consider the conducting function for stormwater collection and absorption effect, in order to optimize the urban water system.

The roads, sidewalks, leisure squares in the residential areas should use permeable paving and sunken green space during the planning and construction process. The sunken green space below the horizon level could store, harvest, retain & detain stormwater more efficiently, and contribute as a part of hydraulic landscape in the residential zones. It's a smart move to cover the architects with green roofs. This practice harvests and retains stormwater,as well as provides the city with a pleasing landscape and mitigate urban heat island effect.

(3) To establish a assurance mechanism of Sponge City management implementation.

We should promote the standardization and legalization of Sponge City mainly through lawful establishment, institutional reform, regulatory and supervisory management refinement and other measures alike. First, to establish and improve the legal & policy system for Sponge City,and provide an institutional guarantee for Sponge City construction. Second, local governments should reform the current urban stormwater management system, including the establishment of the reward & punishment mechanism associated with the performance assessment and the investment & financial system dedicated in Sponge City construction ,and change the current opaque urban plan condition result from lack of social supervision management system. The public supervision from society should be strengthened throughout the process of sponge city construction no matter in which phase, such as project approval and initiation, planning and design, engineering and construction, etc.

The philosophy of Sponge City brings the city back to nature. In this era, nature and humans is an entity. The future is the alternative result of our choice, not our inevitable destiny. Imagine, when the year of 2200 or 2500 approach, our descendent may compare us to aliens who only treated earth as a stopover for refueling. Moreover, our offspring may regard us as barbarians who have destroyed their own homeland. Contemporarily, we are supposed to form a sharing atmosphere instead of an unlimited plundering culture where all the creatures on earth could popularize and generate together. As a newly emerging urban development model, Sponge City theory is still in the stage of exploration and development. We are urged to perceive its concept and content in a broaden horizon and deeper dimension. The practice strategies and standards should be developed and formulated in line with our countries' specific geographical characteristics, catalyze and stimulate the improvement and maturity of the theoretical system, facilitate and prompt cities back to nature. In a nutshell, the outlook of our task is promising but far more than smooth.

He Fang
Sep.2015
(Translated by Zhang Xu)

CONTENTS 目录

规划尺度下的雨洪管理
STORMWATER MANAGEMENT IN PLANNING SCALE

014 构建区域绿色基础设施,走向精明规划设计—— 规划尺度下的雨洪管理
BUILT UP THE REGIONAL GREEN INFRASTRUCTURE, CARRY ON THE SMART PLAN AND DESIGN - STORMWATER MANAGEMENT IN THE URBAN PLANNING SCALES

018 ACTIVATE A RESILIENT RIVER: MINNEAPOLIS WATERFRONT CITY DESIGN
弹性的河流:明尼阿波利斯滨水城市设计

028 THE COMPREHENSIVE IMPROVEMENT OF THE LANDSCAPE AND WATER SYSTEM OF SHENZHEN FUTIAN RIVER AND CENTRAL PARK
深圳福田河与中心公园生态景观及水系综合整治

036 SLOW DOWN: LIUPANSHUI MINGHU WETLAND PARK
让水流慢下来:六盘水明湖湿地公园

044 MINGCUI LAKE LANDSCAPE DESIGN, LAKESIDE NEW ZONE, HEKOU DISTRICT, DONGYING
东营市河口区湖滨新区鸣翠湖景观设计

052 URBAN DRAINAGE & WATERLOGGING PREVENTING PLANNING, COPENHAGEN
哥本哈根城市排水防涝规划

060 A GREEN SPONGE FOR A WATER-RESILIENT CITY: QUNLI STORMWATER PARK, HARBIN
绿色海绵般的水适应城市:哈尔滨群力雨洪公园

070 RE-BORN OF NATURE AND CULTURE: RE-ESTABLISHING THE DELTA IN A NEW CITY
自然与文化的重生:在新城中重塑三角洲区域

078 WATER CORRIDOR OF SHENZHEN UNIVERSIADE CENTER
深圳大运中心水廊道

086 FLOATING CONNECTION: HARBIN CULTURAL CENTER WETLAND PARK
漂浮的连接:哈尔滨文化中心湿地公园

096 RESILIENT LANDSCAPE: JINHUA YANWEIZHOU PARK
弹性景观:金华燕尾洲公园

106 NANHU: REIMAGINING THE AGRICULTURAL VILLAGE
南湖:农业村落的重塑再造

114 OFFENBACH HARBOUR, PLANNING
奥芬巴赫港口区域规划

120 LANDSCAPE PLANNING OF FUYANG WEST RIVERFRONT
富阳北支江滨江区景观规划

126 OCT HAPPY COAST IN SHENZHEN
深圳华侨城欢乐海岸

134 URBAN GREEN VALLEY LANDSCAPE PLANNING & DESIGN IN NORTH STATION CENTRAL BUSINESS DISTRICT, SHENZHEN
深圳北站商务中心区城市绿谷景观规划设计

142 NINGBO ECO-CORRIDOR
宁波生态廊道

154 WALLER CREEK: CITY OASIS
沃勒溪:城市绿洲

164 A GREEN TIE AT THE CENTER OF THE NEW CITY: SUSTAINABLE DESIGN OF SHANGHAI JIADING NEW CITY LANDSCAPE AXIS
上海嘉定新城中央绿色纽带:"紫气东来"可持续设计

170 WUSONG RIVERFRONT WATER TREATMENT PARK
吴淞滨江净水公园

178 SHENZHEN SOIL AND WATER CONSERVATION PARK
深圳市水土保持科技示范园

186 LANDSCAPE PLANNING OF CHANGSHA BAXI ISLAND
长沙巴溪洲景观规划

196 ECOLOGICAL DRAINAGE & WATERLOGGING PREVENTING SYSTEM IN CULTURAL CENTER, TIANJIN
天津文化中心生态排水防涝系统设计

204 CONCEPT AND PRACTICE OF THE ECOLOGICAL CAMPUS DESIGN: LIAONING POLICE AND JUDICAL MANAGEMENT CADRES COLLEGE NEW CAMPUS DESIGN
生态校园的综合设计理念与实践：辽宁公安司法管理干部学院新校区设计

214 CREATING HABITAT, COLLECTING RAINWATER, MAKING BIOLOGICAL REPRODUCTION: ECOLOGY PARK
引生境、承天露、生万物：深圳湾科技生态园景观设计

景观设计尺度下的雨洪管理
STORMWATER MANAGEMENT IN LANDSCAPE DESIGN SCALE

226 宜居城市下的水敏性生态系统
WATER SENSITIVE ECOLOGICAL SYSTEM IN THE LIVABLE CITY

228 LINEAR PARK AND WATERWAY OF RAYCOM CITY, HEFEI
合肥融科城带状公园与水道设计

236 GUTHRIE URBAN GREEN GARDEN
格思里城市绿地

242 ZOLLHALLEN PLAZA, FREIBURG
弗莱堡市扎哈伦广场

248 SEATTLE CULTURE SQUARE
西雅图文化广场

254 DAYLIGHTING OF ALNA RIVER: HOLALLOKA CENTRAL PARK, OSLO
埃纳河的日光：奥斯陆海伦拉卡中心公园

260 LANDSCAPE PLANNING AND DESIGN OF GIANT INTERACTIVE GROUP
巨人网络集团总部景观规划设计

270 URBAN PLANNING AND STORMWATER MANAGEMENT OF GAIIOSMANPASA, ISTANBUL
伊斯坦布尔格基奥马帕萨区城市规划和雨洪管理

278 CALIFORNIA ACADEMY OF SCIENCES
加州科学博物馆

284 WATER MANAGEMENT IN MCLAREN TECHNOLOGY CENTER, LONDON
伦敦麦克拉伦技术中心水管理项目

290 CHINESE SQUARE & ROUND
中式"方圆"

296 LANDSCAPE DESIGN OF SHENGYIN YUAN, TSINGHUA UNIVERSITY
清华大学胜因院景观设计

304 DESIGN OF UP+SDESIGN RAIN GARDEN OF 768 CREATIVE INDUSTRIAL PARK
768创意产业园区阿普雨水花园设计

新加坡 ABC 导则
SINGAPORE ABC GUIDELINE **310**

SPONGE CITY

规划尺度下的雨洪管理
STORMWATER MANAGEMENT IN PLANNING SCALE

013

014

构建区域绿色基础设施，走向精明规划设计
——规划尺度下的雨洪管理

近数十年来，城市雨洪内涝灾害愈演愈烈，且在世界各地均频繁发生。除去自然因素，人为活动及其导致的水文功能改变则是关键原因。一方面，现代城市建设导致各类城市硬化地表、人工设施增加，使城市综合径流系数提高，径流总量大为增加。另一方面，高强度人为活动干扰已强烈改变了自然水循环，在一定时空范围内超出了城市系统的调控能力。如有限的雨水管井难以承受短时峰值径流骤增，城市河湖水域面积的减少压缩了水的产汇流路径及缓冲空间，大面积连续硬化下垫面阻断径流下渗导致大规模局部积水，单一快排改变了水体下泄的时空强度与容量，调水、渠系、管网等人工路由极大改变了自然产流、汇流、径流、下渗过程等。

人类目前的城市雨水处理主要采用人为工程化措施，即所谓"灰色基础设施途径（Grey infrastructure）"，包括从明渠、暗沟、合流/分流制地下管网到堤防、泵站、闸控、水库及调蓄深隧等一系列工程，其目标是快排。然而，国际上不断发展的 BMPs、LID、WSUD、GSI 等理念，均强调自然生态化的雨水处理，充分利用场地内各类自然要素，即绿地、水体、湿地及荒地、农田等，通过土壤、植物等实现对雨水径流的下渗、滞留、调蓄、净化、利用，维持开发建设前的场地水循环及径流水平，从而解决雨洪问题，这可称为"绿色基础设施途径（Green infrastructure）"。

从人居环境及人类生存发展的需求出发，二元水循环、灰绿础设施皆须兼顾。只是长期以来，人们更关注人工水循环，为了自身需求，对水资源一味索取，一意改变，而面对水灾害，也一直依赖"灰色基础设施"以求安全。然而，现在及未来的重点，却是在以工程技术为主导的高密度城市环境下，尽可能地为自然水循环留足空间，留出路径，在一定尺度上使城市的"人工"与"自然"水循环保持协调与平衡。在多座城市遭受严重内涝灾害之后，中国政府于 2014 年 10 月发布《海绵城市建设技术指南》并启动试点城市申报，使解决城市内涝问题上升到国家行动的层面。"海绵城市"是指城市像海绵一样，在适应环境变化和应对自然灾害等方面具有良好的"弹性"，下雨时吸水、蓄水、渗水、净水，有需要时将蓄存的水"释放"并加以利用。

在宏观层面，海绵城市的建设重点是协调人工建成系统与自然水文系统之间的矛盾，使城市在"改变"与"顺应"自然水循环之间取得平衡，从而实现水安全、水资源、水生态等综合效益。"绿色基础设施途径"无疑是建设"海绵城市"的关键策略。然而，其难点是协调水循环过程与区域土地利用和空间布局之间的矛盾。"绿色基础设施"既可理解为现有的"灰色工程基础设施"的绿色化、生态化、环境友好化、低碳化，如对道路、桥梁、河岸、管线甚至建筑等增加绿色元素、加强生态服务功能。同时绿色基础设施也可理解为连续、系统的"绿色基础空间结构"，包括具有基础结构（Infrastructural）意义的任何绿色元素，如河流、湿地、林地、海岸、公园绿地、山脉等等，重点强调其网络化、多层次、复合性及灵活性、适应性等特点。总之，"绿色基础设施途径"既是一个激发人们反思人地关系的"战略"，具有号召力、整合性与开放性，也是一个指导具体行动的"策略"，具有操作性、科学性与空间维度（如形态、尺度、边界等）。

从雨洪管理功能出发，构建区域绿色基础设施可努力的方向包括：（1）流域治理、生态修复、水源涵养、水土保持及区域水资源综合开发利用；（2）城市水系统规划与河湖生态修复；（3）城市生态排水与雨洪管理系统规划；（4）城市滨水区、港区开发与绿色城市设计；（5）城市湿地、湖泊等水域开放空间景观游憩规划等。

近年来，景观水文（Landscape Hydrology）作为一种具有整合性、创新性的水研究与设计理论日益得到重视。该理论旨在从水科学、生态科学与设计学科等交叉领域中进行一番新的探索方法，并非仅仅是水环境的景观化和美化，还包括从水的"过程-格局-功能"的动态关系入手来研究分析各类水文现象及问题，建立基于科学合理性的整体水设计观，以实现水问题的综合解决，因此可成为构建区域绿色基础设施的指导性理论方法。

因此，"构建区域绿色基础设施，走向精明规划设计"成为规划尺度下的雨洪管理的核心思路。本章的各个国内外优秀案例，从不同的角度、不同的地域及不同的类型，对上述目标、策略与实践作出了充分诠释。它们均体现了对传统绿地系统规划和土地利用规划的突破，一方面超越了传统绿地的空间范畴，在更大范围内从多尺度、多层次、多功能考虑绿色生态空间的保护与雨洪管理的需求，另一方面在规划理念、方法、技术等方面有所突破，使原本被动留下的"无意义"空白土地成为具有水文与生态过程、功能、结构的文化景观系统，使消极被动保护变为积极主动管理，影响了城市的空间发展模式和建设方式，改变了人类对地球表面空间的利用态度与方式。这与近年国内外涌现的"精明增长""反规划""景观都市主义"等理念皆有内在的联系。

刘海龙

2015 年 9 月

016

Built Up Regional Green Infrastructure, Carry On the Smart Planing and Design – Stormwater Management in Planning Scale

In recent decades, the waterlogging issue result from stormwater became a severe hazard all over the world. Its degree intensified and frequency increased. To a large extent, the human activity is the key reason to the change of hydraulic functions. On the one hand, modern city construction resulted in an increased rate of impervious surface and artificial facilities, making the city's comprehensive runoff coefficient elevated and runoff volume enlarged. On the other hand, the high-intensity of human behavior exerted a strong disturbance on natural water cycle, beyond the regulatory capacity of the urban system in a certain time & space range. Namely, the limited rainfall tube-wells are vulnerable when the runoff volume climax to the peak in a short time. The reduced urban river & lake area would compress water convergence path and buffer space. A large area of continuous hardened underlying surface blockade the runoff infiltration process and result in a large scaled partial ponding. The simple & fast discharge pattern has changed the intensity and capacity of water drainage in the dimension of time & space. Water diversion canal, channel & pipeline and other manual routing disturbed the natural process considerably, such as runoff production, runoff gathering, runoff inflow, runoff infiltration, etc.

The current stormwater treatment strategy is mainly relied on the artificial engineering measures, the so-called gray infrastructure approach. (Gray Infrastructure) These measures are featured by a series of construction work ranged from the open channels, culverts, confluent/diffluent underground pipe network systems to the dikes, pumping stations, gated buffers, reservoirs and regulatory deep tunnels. All these serve for the only goal of quick discharge. But the continuously refining international concepts like BMPs, LID, WSUD, GSI, etc., all have emphasized on natural & ecological stormwater treatment strategy and highlighted in fully using all kinds of natural elements on site, particularly, green space, water body, wetland and wasteland, farmland, etc. To achieve runoff infiltration, retention, regulation, storage, purification and utilization through the soil and plants, and to tackle the stormwater issue in this manner, all these could be called green infrastructure approach. (Green Infrastructure)

Set out from the human settlement-environment demand of existence and development, binary water circulation and gray are both required to be taken into account. But in order to meet our own needs, people are more concerned about the artificial circulation for a long time. When confront with water source, only ask for, not contribute in, only alter it, not adapt to it. And in the face of water-related disasters, people are excessively depended on the gray infrastructure to ensure the security. The crucial task at present and in the future focuses on leaving enough space for the natural hydraulic cycle and path. Although the urban environment of high-density is dominant by artificial engineering & technical construction work, we should try our best to balance and coordinate the relation between artificial water circle and natural water circle. China published Technical Guide of the Sponge City Construction in October 2014 after Chinese cities suffered seriously from waterlogging, Started up with the declaration of pilot cities, and brought the waterlogging problem into the public concern and advocated nationwide action. The concept of Sponges City means the city can act like a sponge, to be resilient and flexible enough to adapt to the climate change and natural hazards. The city can absord water, store water, infiltrate water, purify water when it rains,while release the conversed water and utilize the water when it in need.

In a macro view of perception, the fundamental objective of Sponge City construction is to coordinate the contradiction between artificial water system and natural hydrological network, to reach to a balanced situation between altering natural water cycle and conforming to it, in this way to obtain the comprehensive benefits from water security, water resources and water ecology. Green infrastructure approach, undoubtedly, is the key strategy to build Sponge City. But it's difficult to coordinate the conflict among the water recycling process, regional land use and space layout. For green infrastructure, it can be regard as a greening, ecological, environment-friendly, and low-carbonized process towards the current grey engineering infrastructure. For instance, attach and strengthen the green elements' ecological-service function to the roads, bridges, riverbanks, pipelines, and even buildings. It is also refer to the continuous & systematic basic green spatial structure, include all kinds of green factors characterized by the infrastructural (fundamental and structural) significance, just like rivers, wetlands, woodlands, coastal banks, green parks, mountains and so on. They are all multi-level, complex, flexible, resilient and adaptable. In general, green infrastructure approach is an appealing, integrated and open strategy, which can inspire people to reflect on the relationship between people and land. But also a feasible and scientific strategy, which can guide the specific action and practice in the spatial dimensions (such as shape, size, boundary, etc.

In the terms of storewater management function, the direction of green infrastructure construction we could put effort in includes: (1) watershed management, ecological restoration, water conservation, soil & water conservation and comprehensive utilization of regional water resources (2) urban water system planning and lakes &rivers ecological restoration (3) urban ecological drainage and stormwater management system planning; (4) City waterfront & port area development and green city design; (5) urban wetlands, lakes and other open water space recreational landscape planning, etc.

In recent years, as a kind of integrated and innovative water study and design theory, landscape hydrology has gained a wide attention. The theory aims in discovering more knowledge and methods from the intersection of water science, ecological science, design and other disciplines. Instead of simply realize the embellishment & beautification of the landscape, the study dedicated in analyzing various types of hydrological phenomena & problems through the dynamic water relation mode of process-structure-function. We should establish a rational science-based water master design philosophy, in order to accomplish a comprehensive strategic solution to the water problem. In this way, the theory can be used as the guiding methods in the regional green infrastructure construction.

Therefore, the core idea of stormwater management in the urban plan scale was replaced by built up the regional green infrastructure, and move towards the smart plan &design. The outstanding cases over seas and at home in this chapter, illustrated &interpreted the above-mentioned objectives, strategies and practices incisively in the aspects of disparate visual angles, different regions and various types. They all reflected a breakthrough in the traditional green space system planning and land use planning. On the one hand, these cases have blurred the boundary of traditional green space. It considers the green ecological space protection and e stormwater management demand from a multi-scale, multi-level and multi-functioned view. On the other hand, these cases made a breakthrough in the planning concept, methods and techniques. Transform a passively remained meaningless blank-land into a cultural landscape system with its own hydraulic and ecological process, function and structure. Apart from this, negative protection &conservation, was replaced by positive management & practice. The impact on spatial development patterns and urban construction methods, have changed human' attitudes and behaving manners toward the surface space utilization on earth. There's an intrinsic link between the newly-emerged concepts domestically and internationally in recent years, namely, smart growth, anti-planning, landscape urbanism, etc.

Liu Hailong

Sep.2015

(Translated by ZhangXu)

018

弹性的河流：明尼阿波利斯滨水城市设计

ACTIVATE A RESILIENT RIVER: MINNEAPOLIS WATERFRONT CITY DESIGN

项目位置：美国明尼苏达州明尼阿波利斯市

项目规模：8.8 km

设计公司：土人景观

Location: Minneapolis, Minnesota, USA

Size: 8.85 km

Design Firm: Turenscape

明尼阿波利斯市公园与游憩委员会和公园基金会共同主办了一项设计竞赛。由土人景观领衔并由其他6个设计公司参与合作，提出了沿密西西比河水道8.8 km（5.5 miles）的滨水带设计方案。场地范围从石拱桥至城市北段，目前这一区域属于"被遗忘"的地区，场地现有土地使用功能包括住宅、工业区、融合多个种群及文化背景的社区、所有权归政府的公园，以及历史街区等。土人团队提议把景观作为生态基础设施和一种综合性的工具来应对生态、社会、经济和文化的挑战，为目前被忽视的场地勾勒了21世纪新型景观和城市肌理的蓝图（经未来数十年才能完成）。

挑战与目标

自1867年明尼阿波利斯市成立以来，密西西比河一直是城市发展的动力。城市及其滨水地区从开始的木材厂到后来的面粉厂拥有非常丰厚的工业底蕴。尽管城市下游滨水区的发展取得巨大的成功，但上游地区亟待开发和设计。该地区所面临的挑战包括产业用地不断缩减（损失了数千米）、原有大尺度基础设施将河流和周围的社区割裂开来以及公共绿地不足等。

设计团队面临以下四项挑战。

（1）生态修复：如何设计上游的滨水地区和周围的社区，以重建健康的自然生态系统并最大限度地利用公园系统的生产力？如何适应未来全球气候变化的影响？如何发现一种能够反映并预测未来生活方式的新美学标准，且在这种生活方式中，文化和人类活动过程能够适应自然环境的变化，并对其进行调解？

（2）社会公平：上游滨水区的社区，尤其是明尼阿波利斯北部地区的公园用地或其他休闲设施非常少。这种"不平等"加剧了其他社会挑战。那么如何通过该项目营造更加公平的社会环境，使明尼阿波利斯北部像该城市其他地区一样成为城市居民乐于前往的目的地？

（3）激活经济：上游滨水区的工厂和商业区都背对壮观的密西西比河。如何在恢复活力的河流廊道中加快发展相关的新型产业？如何使投入的资金在未来催生更广泛的经济活动和商业活动？

（4）文化认同：明尼阿波利斯地区曾是美国土著居民重要的集聚地。随后，它成为欧洲移民者的目的地，他们在此从事毛皮贸易、木材加工和粮食储藏。（例如：对木材历史的记忆可通过公园设计中木材的使用来表达；砂石堆料和大型机械可保留在公园中）一个新的河岸或许能让不同阶层和不同种族的人团结在一起。

设计方案

设计团队提出了以下三个应对挑战的策略。

（1）建设生态基础设施：强大的城市公园体系扎根于强健且可修复的大自然中。找出可利用的自然资源，设计一个保护和加强自然和文化过程的生态网络，这些过程包括绿色交通、自然雨水处理、城市农业和其他绿色基础设施。

（2）使城市生活回归河流：随着生态基础设施的发展，将河流等自然资源重新定位，将学校和住宅、办公和科研、艺术和商业在内的功能都转移到河边。

（3）在发展中调整蓝图：城市领导人认识到虽然城市的土地用途和建筑不断发生变化，但基本的景观元素继续存在——包括河流和周边地形。这一规划和设计憧憬了未来50年的城市理想图景，基于这些基础生态资产，打造连续的步行和自行车通道，鼓励园艺，建设文化艺术设施，完善绿色交通网络等等，让人们重新回归河流。

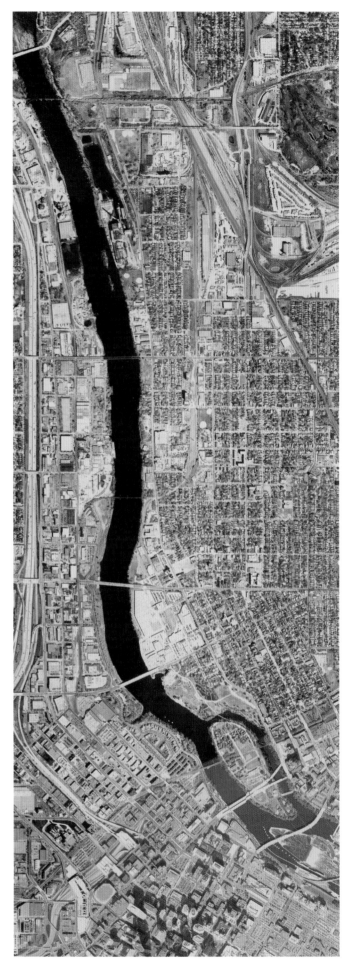

规划区域的现状航拍照片
Aerial view of planning sit

In a competition sponsored by the Minneapolis Park & Recreation Board and the Minneapolis Parks Foundation, Turenscape teamed with six other firms and developed this proposal for 8.8km (5.5 miles) of waterfront redevelopment along the Mississippi River. The currently neglected site, which spans from the Stone Arch Bridge to the flatlands of the northern city limits, encompasses a variety of land uses from residential to industrial, neighborhoods of demographic and cultural diversity, government owned parks, and historic districts. The team proposed using landscape as ecological infrastructure and an integrative tool to address ecological, social, economic, and cultural challenges, and creating a 21st-century vision (to be implemented over decades) for a new landscape and urban fabric for this now neglected site.

Challenges and Objectives

Since Minneapolis's founding in 1867, the Mississippi River has been the engine of its development. The city and its riverfront have a rich industrial past, starting with lumber mills and then flour mills. Whereas the city's lower riverfront can boast tremendous contemporary success, the upper riverfront is waiting to be rediscovered and redesigned. Challenges facing this area include a diminishing industrial base (loss of mills) and large-scale infrastructure that divides the river from surrounding communities, as well as public parks that are too small for their constituent communities.

The design team has identified four primary challenges.

(1) Ecological renewal: what can be done on the upper riverfront and in its surrounding communities to rebuild a healthy natural ecosystem and make best use of a park system for its productive potential? How can the site adapt to the coming effects of global climate change? How can we discover a new aesthetic that reflects and foresees a lifestyle in which cultural and human processes adapt to and mediate the changing natural environment?

(2) Social equity: Most Minneapolis communities of the upper riverfront have very little parkland or other public amenities. This inequity compounds and exacerbates other social challenges. How can this project create a more equitable society in which North Minneapolis becomes as much of a destination for the whole city as its other, wealthier areas?

(3) Vibrant economy: Commercial buildings like factories and warehouses in the upper riverfront include many that present blank walls to the majestic river. How can we foster new better related industries in a resurgent river corridor? How can investments in the river catalyze broader economic activity and attract more businesses of the future?

(4) Cultural identity: The site of Minneapolis was an important gathering spot for native USAns. It subsequently became home to European immigrants and an active base of fur trade, lumber milling, and grain storage. This history should be noted and its remnants preserved in any new development (e.g., the history of logging can be recalled in the use of large logs in park designs; gravel production aggregate piles and large machinery can be incorporated into parks). A new riverside may also contribute to bringing people of diverse classes and ethnicities together.

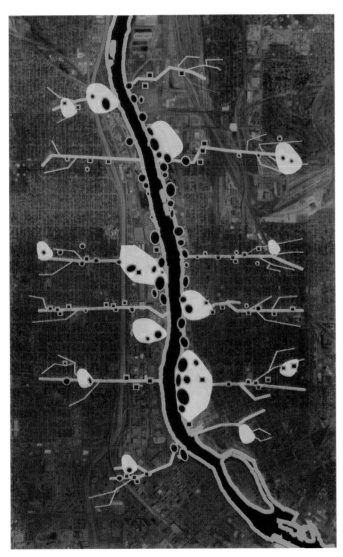

建设生态基础设施,包括绿道、湿地、农耕用地以及雨洪管理系统。
Build ecological infrastructure, including green corridors, wetlands, agricultural fields, and stormwater management.

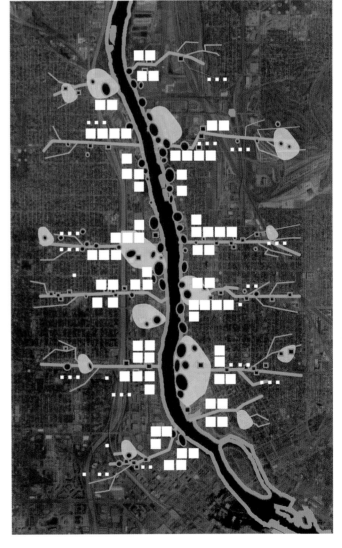

重建城市与水的关系:将建筑与生态基础设施相融合。
Reorient urbanism to the river: Align new buildings with green infrastructure.

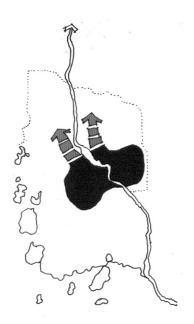 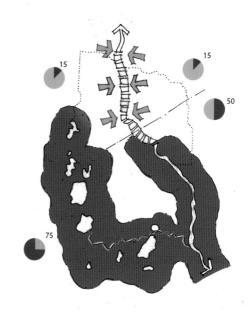

Design Strategies

The design team has proposed three strategies to address these challenges.

(1) Build an ecological infrastructure: A strong city park system is rooted in robust and resilient nature. Natural assets were identified to help to design an ecological network that protects and enhances critical natural and cultural processes. These include sustainable transportation, natural storm water management, urban agriculture, corridors for wild flora and fauna, wetlands and other green infrastructures.

(2) Reorient urbanism to the river: As green infrastructure grows, the city can be reoriented toward the river, with schools and housing, jobs and research, art and commerce placed alongside the new ecological infrastructure. Skywalks and linear parks can connect residential centers to the riverside.

(3) Curate the vision through time: The city's civic leaders have understood that its land uses and architecture will change, but that its fundamental landscape elements—the river and surrounding topography—will endure. This planning and design scheme explores ideal growth over a fifty-year period, grounded in those essential natural elements, reconnecting people to the river through unobstructed movement paths, creating riverfront horticultural and cultural destinations, building transportation networks to the river, and more.

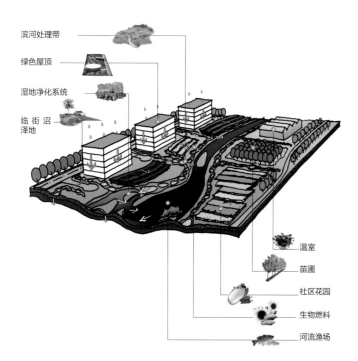 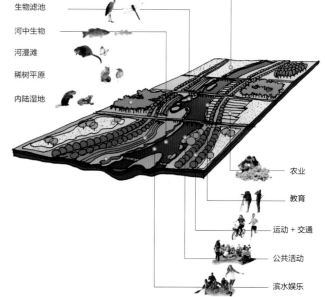

设计自然排水系统以管理雨水。多种生态过滤系统被用于过滤来自周边邻里、84号州际高速公路和河流阶地上，居民区排放的生活污染物。在河滩地上增加滞水空间；在上游河流阶地上，利用疏浚河道过程中挖出的沙石堆积起伏有致的独特景观。在私人花园、社区农田、花圃和苗圃地之间，分布着野生动物栖息地。

A system of natural drainage is proposed to manage water of flows. Various natural cleansing systems will be used to treat polluted runoff from surrounding neighborhoods, interstate highway 84, and the inhabited river terraces. Retention space is to be increased in the flood plain, and excavated materials will be deposited in the upper river terraces to create dramatic landforms and overlooks. Wildlife habitat will be interspersed with personal allotment gardens, community-supported agriculture, along with aquiculture and tree nurseries.

建设生态走廊，以提高环境的生物多样性。密西西比河是鸟类迁徙的重要廊道。多种乡土植物和作物将为鸟类、鱼类和其他野生动物提供食物和栖息地。

The ecological corridors are built to increase habitat biodiversity. The Mississippi is a flyway for migrating birds. A palette of native plants and crops will provide food and shelter for birds, fish, and other wildlife.

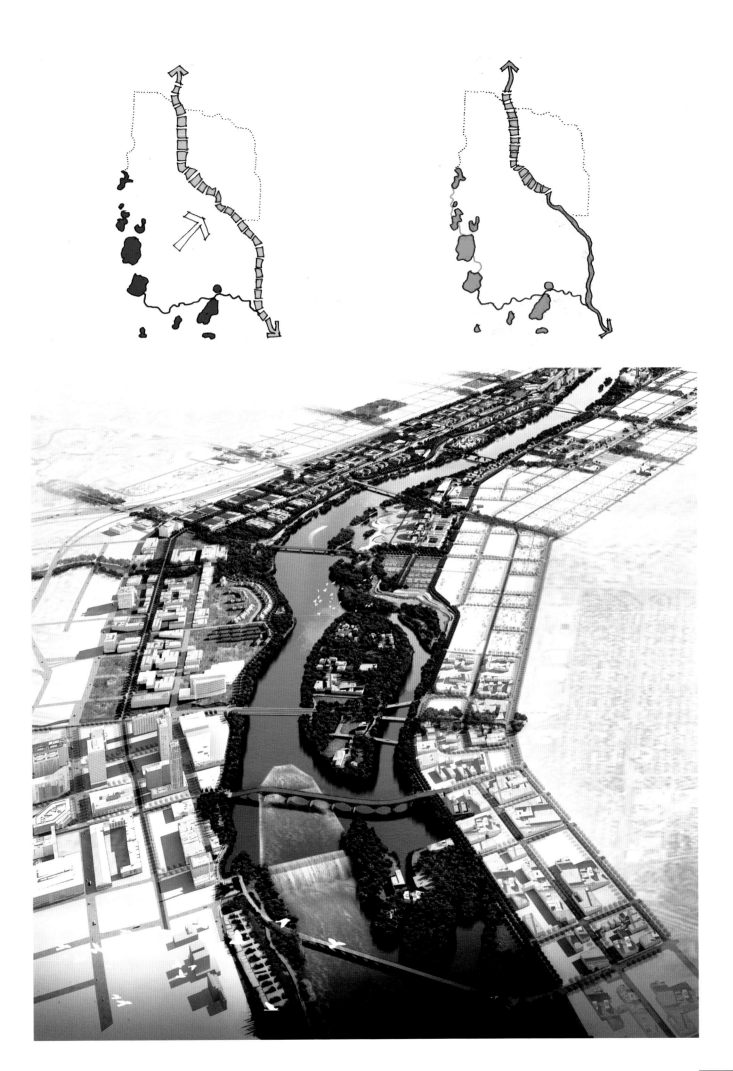

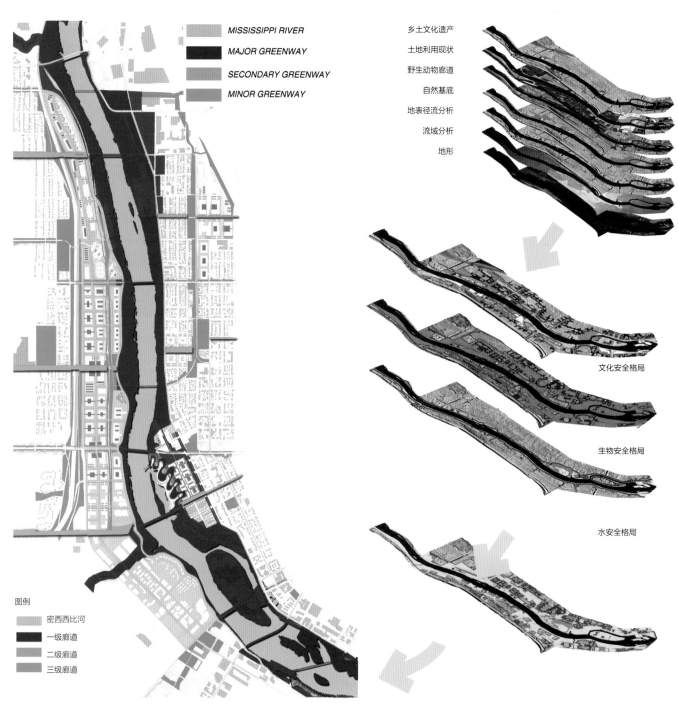

生态基础设施网络

生态安全格局

战略性的景观格局： 构建景观安全格局，以高效地维护生态过程。将水文、生物和文化遗产的保护等各种过程的景观安全格局进行叠加，以此建设生态基础设施。这一综合性的格局就像树一样：树干为密西西比河道，绿色的树枝延伸入城市肌理内部，树叶则为具体的场地和邻接的功能节点。

Take a strategic conservation approach: "Security Pattern" that offer the most effective ecological protection for the landscape were identified. This was envisioned as a vertical stack of interdependent layers that protect hydrological and biological processes, as well as the cultural heritage. Its network resembles a tree: The river is its truck; branches of green link it to the urban fabric; twigs are the specific sites and features of the neighborhood.

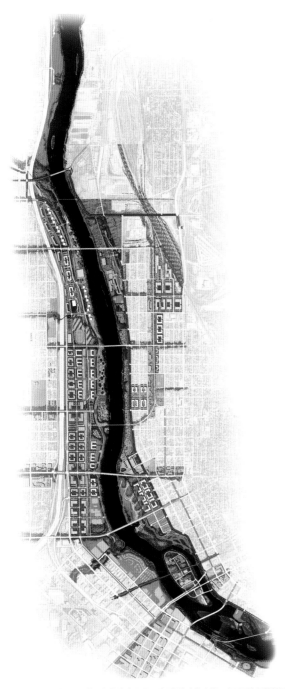

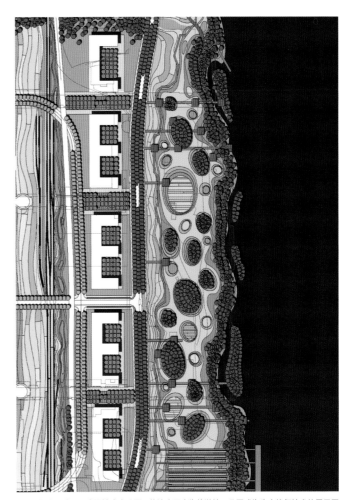

5到10年内:建设一个湿地生态公园,将滨水区变为旅游地。公园成为生态修复技术的展示平台。湿地净化从密西西比河引入的水以及周边社区排出的地表径流。在公园中建设高草覆盖的小岛,成为候鸟和其他野生动物的滨河栖息地,而其建筑材料都来自场地上材料的再利用和原有木料粉碎场材料的有机合成。

Years five to ten: Create a riverfront destination with a wetland eco-lab park. This This large park will demonstrate ecological restoration techniques. The wetlands will cleanse water drawn from the Mississippi, as well as surface runoff from the neighborhoods. Islands for wet prairie and riparian habitats for migrating birds and other wildlife will be built from recycled demolition materials and organic material from the soil factory.

50年后的密西西比河和上游理想的滨水景观:赋予场地文化功能。密西西比河曾经是明尼阿波利斯市的重要休憩资源,长岛的格柏浴场和砾石溪流的韦伯池曾经是居民向往的地方。在此可进行垂钓、划船、游泳、滑冰、桑拿和其他亲水活动。

Vision of an ideal river landscape fifty years hence: Cultural uses are proposed: The Mississippi was once a significant recreational resource for Minneapolis, with community destinations like the Gerber Baths at Halls Island and Webber Pool at Shingle Creek. The river will again host opportunities for fishing, boating, swimming, ice skating, sauna, and other water-based experiences.

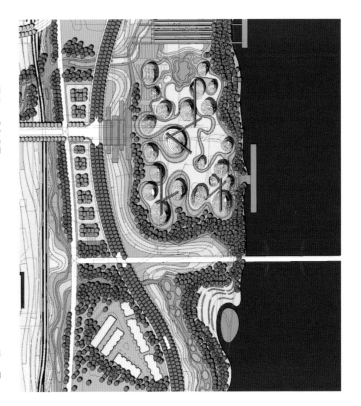

5到10年内:在原河砂堆放场上,建成一个艺术公园和露天剧院,将滨水区变为旅游目的地。湿地净化后的水向南流进公园。

Years five to ten: Create a riverfront destination with aggregate art park and amphitheater. Clean water from the wetlands will flow south into this park.

蜿蜒的沟渠在以后工业时代艺术品和乡土草原为背景的水渠中穿梭。水渠在冬季可成为滑冰道。不远处的露天剧场可举办演出，同时，也可俯瞰中心城区的美景。

Sinuous canals will allow water play among post-industrial artifacts and aggregate piles between ribbons of prairie. The canals will become circuits for ice skating in the winter. With a great view to downtown, an adjacent amphitheater will host performing arts.

湿地公园效果图（观鸟亭和栈道引领游人进入部分区域）

Picture of wetland park (A viewing platform and a network of boardwalks will allow partial access to the wetlands)

10到15年内：建设环线公交轻轨交通，延伸公园路，催化河流上游滨水地区产业和社区的形成。公交轻轨线延伸了城市的公共交通系统，以促进城市发展。改造利用现有的铁路货运桥，使其成为轻轨交通和步行桥，连接河两岸。步行者和骑自行车的人可尽情享用河两岸连续的公园路。

Years ten to fifteen: Build transportation infrastructure with a streetcar loop and extensions of the parkways for new industries and new communities to emerge along the upper riverfront. The streetcar loop will extend the city's transit network to catalyze development. A transformed BN bridge will bring the streetcar and pedestrians to both sides of the river. Walkers and bikers will enjoy an uninterrupted parkway network along the river.

现状：铁路货运桥，对过河的人来说缺乏吸引力

The existing BN bridge makes it unattractive to cross the river on foot

028

深圳福田河与中心公园生态景观及水系综合整治

THE COMPREHENSIVE IMPROVEMENT OF THE LANDSCAPE AND WATER SYSTEM OF SHENZHEN FUTIAN RIVER AND CENTRAL PARK

项目位置：中国广东省深圳市
项目规模：1590 ha
设计公司：深圳市北林苑景观及建筑规划设计院、深圳市水务规划设计院

Location: Shenzhen, Guangdong Province, China
Size: 1,590 ha
Design Firm: Shenzhen BLY Landscape & Architecture Planning & Design Institute, Shenzhen Water Planning & Design Institute

总平面图
site plan

福田河为深港界河深圳河的支流，位于深圳市中心区，北起梅林坳，依次穿越梅林、笔架山公园、中心公园，沿福田南路在皇岗口岸东部汇入深圳河。河道流域面积15.9 km²，主干流全长6.8 km。改造前河道防洪能力较低，且河槽为生硬的"三面光"水泥工程做法，既不生态，也与周边环境尤其与流经的公园相分离。同时，部分河道护砌损坏严重，水质发黑污染严重且生态性、景观性、亲水性差。

2005年，面对福田河存在的诸多问题，同时在对中心公园历史保留的大片果林的改造之际，市政府决定对福田河进行全面、系统地综合治理，解决福田河水污染严重、防洪能力低、生态景观不佳等问题，使其成为中心公园有机的组成部分，营造优美的自然水岸风景。

福田河的整治以生态治理为指导思想，以满足河道防洪要求为前提，通过截排污水和初期雨水、利用再生水补水、中水再净化、坡岸生态覆绿等措施，强化雨水蓄积与下渗，缓解洪涝危害。同时，改善河流水质，恢复河道生态景观功能，使生态与防洪两者兼顾，将"海绵城市"的设计理念充分运用到整个河道建设中。通过打造宜人的滨水休闲空间，使福田河成为中心公园的生态水景，为市民营造亲水、赏水、玩水的环境，满足市民亲近自然与赏景游憩的需求。

提高河道防洪能力

基于福田河重要的地理位置，规划通过在河道流经的中心公园园区内设置一定面积的滞洪区来解除洪水对福田河沿线城区的威胁，将防洪能力提高到百年一遇的水平。

滞洪区的设置结合公园的总体规划，迁走原有果林，开挖地形形成低洼地，洼地的一部分成为公园的疏林草地，另一部分再挖深成为公园的湖泊，这样不仅可解除百年一遇的洪水威胁，又可成为公园的湿地景观，使"海绵城市"的技术理念得以充分体现。

改善河道水质

福田河采用"初雨收集管涵+分散调蓄池+河道污水泵站或初雨抽排泵站"的方案，对河道进行污水截流，保证河水的水质。通过管涵截流两岸难以分流的少量污水并将初期雨水送到滨河污水处理厂，使处理后的中水回补河道，降低河道中的污染物浓度，促进水体交换，增强河道的自净能力。同时，在笋岗西路以南的滞洪湖泊中增加湿地生态岛，再次净化补入河道的中水，使水质达到景观用水的标准，保证河道的水质与水源满足河道的观赏性水体要求。

恢复自然生态水岸

为恢复河流的自然生态属性，对坡岸和河床进行生态景观改造。在满足河流的防洪功能的同时，改变原有线形单一的河道岸线，丰富河道断面，形成浅滩和深水，营造生物多样性的环境条件；拆除已有浆砌石或混凝土硬质护底，并将部分拆除的石头处理后干摆于河床，使雨水下渗，加强与周围环境的交换，营造流态的多样性，构筑河床水生动物的多元化栖息空间。

福田河综合整治是深圳市政府实施的一项民心工程，也是深圳市水环境整治重点工程之一。它不仅使河道的生态景观功能得以全面恢复，回归自然的雨水蓄积与渗透，重构"海绵基底"，还有效地将河道的防洪标准提高到百年一遇的水平。南北纵贯笔架山公园与中心公园的福田河，成为联系两大公园的纽带，共同构建城市中心区的生态景观走廊。

如今，福田河沿线波光潋滟，流水潺潺，草木葱葱，鸟语花香；昔日令人掩鼻的"臭水沟"已水清岸绿，重现自然生机。

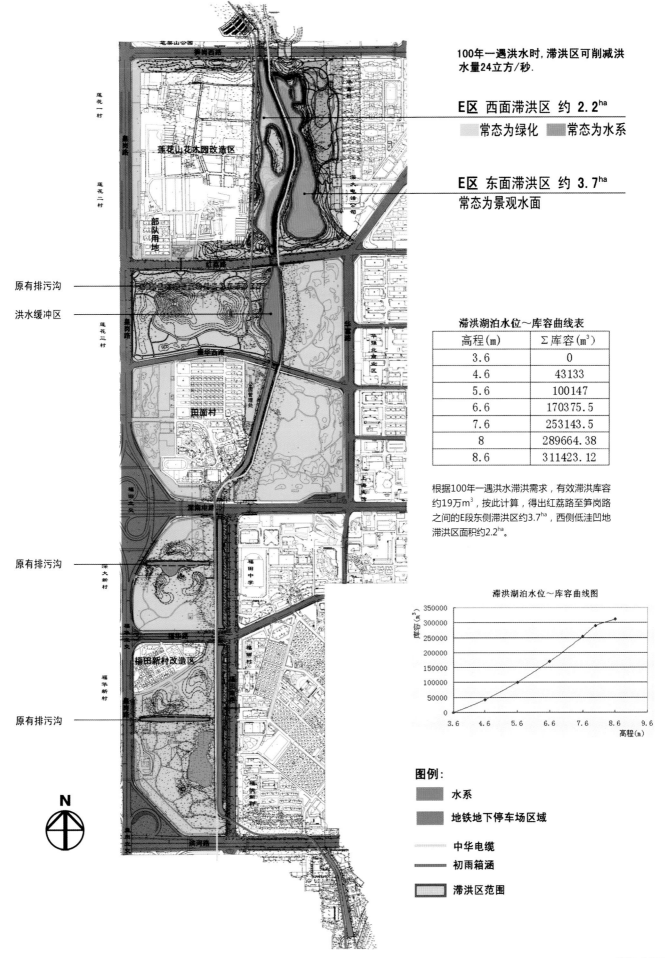

100年一遇洪水时，滞洪区可削减洪水量24立方/秒.

E区 西面滞洪区 约 2.2ha
■ 常态为绿化　■ 常态为水系

E区 东面滞洪区 约 3.7ha
常态为景观水面

滞洪湖泊水位～库容曲线表

高程(m)	Σ库容(m³)
3.6	0
4.6	43133
5.6	100147
6.6	170375.5
7.6	253143.5
8	289664.38
8.6	311423.12

根据100年一遇洪水滞洪需求，有效滞洪库容约19万m³，按此计算，得出红荔路至笋岗路之间的E段东侧滞洪区约3.7ha，西侧低洼凹地滞洪区面积约2.2ha。

滞洪湖泊水位～库容曲线图

图例：
■ 水系
■ 地铁地下停车场区域
⋯⋯ 中华电缆
── 初雨箱涵
■ 滞洪区范围

滞洪区分布图
Layout plan of flood retention areas

Futian River is in the east of CBD of Shenzhen. It is a tributary of Shenzhen River and originates from Meilinao, a hilly area in the north of Shenzhen. Futian River goes across Shangmeilin, Bijiashan Park and Central Park. It goes along Binhe Avenue and flows into Shenzhen River in the east of Huanggang Port. It has a linear flow pattern, with a length of 6.8 kilometers and a watershed area of 15.9 square kilometers. Before renovation, part of the river channel of Futian River has been severely damaged, which leads to poor flood protection and a loss of biodiversity. The water is also badly polluted.

In 2005, the Shenzhen Government decided to launch a program to upgrade FutianRiver, which would help to solve its existing problems and the renovation of the existing orchard in Central Park. The upgrade aimed at improving the water quality and flood prevention of Futian River and increasing its biodiversity, which would create a beautiful riverfront landscape and make Futian River an organic composition of Central Park.

We devise an improvement strategy of "greening and humanization" after taking the ecological management and the flood control function of the river bank into consideration. We manage to restore the ecological diversity by building "talking with the river" waterfront recreation zone, improving the water quality by intercepting sewage, recycling the rainwater and landscape greening. That makes the Futian River a place where citizens can be close to nature.

Improve the flood control capacity of Futian River

Set a certain area of the water attenuation area in the river basin which flows through the center park to remove the threat of flood to the urban areas along the Futian River and will increase to a once-in-a-century flood control ability.

According to the plan, original fruit woods were removed in water attenuation area. Excavation of the terrain formed a bottom land, part of it becomes an open foerest and grassland in the park, and the other part was excavated deeper to form a lake in the park. It can not only control once-in-a-century flood, but also be a wetland landscape in the park, which fully practice the sponge city idea.

Water Purification

The Futian River uses the facilities of rain water collecting pipe culvert, the storage pool, the sewage pumping station or the rain water drainage pump, so that the sewage of the river will be intercepted and the water quality is guaranteed. The remaining sewage that is hard to distributed and the initial rainfall will be transferred to the sewage disposal plant. After being purified, the reclaimed water will go back to the river to decrease the pollution level, encourage the water exchange and strengthen the self-purification abillity. At the meantime, we build wetland ecological island in the attenuation lakes, which will re-purifying the reclaimed water and make the water quality match the criteria of landscape use.

Recover the natural ecological riverbed

To recover the natural ecology, the bank and riverbed has ecological landscape restoration. While having the flood control function, it also changes the original linear and single river shoreline, rich river section, forming shallow and deepwater and constructing biodiversity habitats. Demolition of Masonry or concrete bottom protection and part of the removed stones is placed in dry riverbed to make rain water infiltration. By strengthen exchange with the surrounding environment, and create a flow pattern of diversity, it builds a diversity of aquatic habitats.

Futian River Revitalization is a popular project implemented by the Shenzhen Municipal Government, and one of the key projects about water environment. It not only full restored ecological landscape features of the river, back to nature and accumulation and infiltration of rainwater, reconstruction "sponge base", and its effective flood control standard will raise the river level to hundred years. North and South running through Bijia Hill Park and Central Park, Futian River, has become a link between the two parks, to jointly build ecological landscape corridor in center area.

Today, Futian River is glittering and gurgling, with lush growth of trees and grass, and birds' twitter and fragrance of flowers. Old "drainage ditch" is green and clear, reproducing natural life.

A段改造效果图
Effect picture of remodification for part A

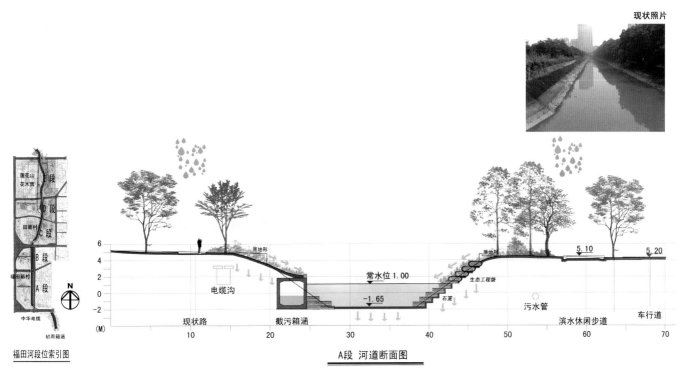

A剖面
Cross Section plan of part A

新增的滞洪区
New flood retention area

灵动的叠水景观
Smart cascading water landscape

充满活力的滚水坝
Dynamic overlow-dam hydrophilic space

湿地生态岛中颇为壮观的美人蕉
Spectacular blooming canna in wetland ecological island

036

让水流慢下来：六盘水明湖湿地公园

SLOW DOWN: LIUPANSHUI MINGHU WETLAND PARK

项目位置：中国贵州省六盘水市

项目规模：90 ha

设计公司：土人景观

获奖信息：2014 美国景观设计师协会设计荣誉奖

2013 中国环境艺术设计金奖

Location: Liupanshui City, Guizhou Province, China

Size: 90 ha

Design Firm: Turenscape

Awards: 2014 ASLA Honor Award in General Design

2013 Gold Medal in Environmental Art and Design of China

明湖湿地公园位于贵州省西部的六盘水市,沿水城河而建。该项目旨在修复河道生态、拓展城市绿色公共空间,同时提升城市河漫滩土地的价值,让湿地成为该市生态基础设施的一部分,并为整个地区的生态服务。经过三年多的设计和修建,明湖湿地公园使原来被污染和渠化的河道恢复了原有的生机,并种植了各类乡土植被,成为整个城市健康生态系统的重要保障。

挑战与目标

六盘水是一个在20世纪60年代中期建立起来的工业城市,以其凉爽的高原气候而著称,城市被石灰岩的山丘环抱,水城河穿城而过。城市人口密集,在60 km²的土地上,居住着约60万的人口。作为改善环境的重要举措之一,市政府委托景观设计师制订一个整体方案,以应对城市所面临的多项挑战,包括以下几个方面。

(1)水污染:作为中国重要的重工业城市之一,六盘水以煤炭、钢铁和水泥行业为主导产业。因此,民众长期受到空气和水污染的困扰。数十年来,从工业烟囱排出的污浊空气中的颗粒物沉积在周边的山坡上,并随着雨水径流被带入河流,来自山坡上农田的化学肥料以及散落的居民点的生活污水也一同随地表雨水径流汇入水体。

(2)洪水和雨涝:由于坐落在山谷之中,该城市在雨季容易受到洪水和涝灾的危害,而由于多孔石灰岩地质,到了旱季又易遭受旱灾。所以,季节性雨水的滞蓄和利用非常重要。

(3)母亲河的修复:20世纪70年代,为了解决泛溢和洪水问题,水城河被水泥渠化。从此,原来蜿蜒曲折的母亲河变成了混凝土结构的、死气沉沉的丑陋河沟,其拦截洪水及环境修复的功能也丧失殆尽。同时,渠化的河道将上游的雨水直泄入下游河道,引发了下游更为严重的洪水问题。

(4)创建公共空间:城市人口激增导致城市休闲和绿色空间的不足。曾经作为城市福音的水系统已经变成城市废弃的后杂院、垃圾场和危险的死角。因此,在人口密集的社区与生态修复后的绿色空间之间建立人行通道极其必要。

这一设计的关键在于减缓来自山坡的水流,建造以自然水过程为核心的生态基础设施,加强雨洪管理,使水成为重建健康生态系统的活化剂并提供自然和文化服务,使这个工业城市变为宜居城市。

设计方案

六盘水明湖湿地公园项目占地90 ha,是该城市规划的综合生态基础设施中首要且至关重要的组成部分。

为了构建完整的生态基础设施,景观设计师同时关注水城河流域和城市本体两方面。首先,河流串联起现存的溪流、坑塘、湿地和低洼地,形成一系列蓄水池和不同承载力的净化湿地,构建了一个完整的雨水管理和生态净化系统,一个绿色海绵体系。这一方法不仅最大限度地减少了城市的雨涝危害,还保证雨季过后仍然有水流不断。第二,拆除渠化河渠的混凝土河堤,重建自然河岸,使河岸恢复生机,使河流的自净能力大大提高。第三,建造包含人行道和自行车道的连续公共空间,增加通往河边的连接通道。这些绿道将城市休憩和生态空间一体化。最后,项目将滨水区开发和河道整治结合在一起。生态基础设施促进了六盘水的城市更新,提高了土地价值,增强了城市活力。

作为六盘水生态基础设施的主要项目之一,明湖湿地公园位于水城河上游区域,设计师面对的是被渠化的河道、被垃圾和污水恶化的湿地区域、废弃的鱼池及管理不善的山坡地,垃圾遍地、污水横流。作为生态基础设施的示范项目,设计的第一步是重建生态健康的土地生命系统,包括改善雨

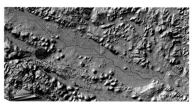

The surface flows of storm water

The regional sotrmwater management system

The concept of the regional ecological infrastructure

The regional ecological infrastructure

场地总平面
Site plan

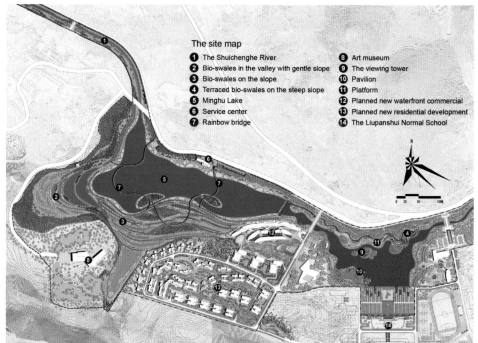

The site map
1. The Shuichenghe River
2. Bio-swales in the valley with gentle slope
3. Bio-swales on the slope
4. Terraced bio-swales on the steep slope
5. Minghu Lake
6. Service center
7. Rainbow bridge
8. Art museum
9. The viewing tower
10. Pavilion
11. Platform
12. Planned new waterfront commercial
13. Planned new residential development
14. The Liupanshui Normal School

Bio-swales on gentle slope

Terraced bio-swales on steep slope

水水质、恢复原生栖息地、建造通向高品质开放空间的游憩道，以促进整个城市的发展。为实现这些目标，具体的工程实施策略包括以下几个方面：

（1）拆除混凝土河堤，恢复滨水生态地带。为各种挺水、浮水和沉水植物提供生境。沿河建造曝气低堰，以增加水体含氧量，促进富营养化的水体被生物所净化。

（2）建造梯田湿地和坡塘系统，以削减洪峰流量，调节季节性雨水。梯田的灵感来源于当地的造田技术，通过拦截和保留水分，使陡峭的坡地成为丰产的土地。它们的方位、形式、深度都依据地质因素和水流分析而设定。根据不同的水质和土壤环境，种植乡土植被（主要采用播种的方式）。这些梯田状栖息地减缓了水流，同时水中过盛的营养物质成为微生物和植物生长所需的养分来源，从而加快了水体营养物质的去除。

（3）人行道和自行车道沿水系铺展，在湿地梯田之间形成网络。大量座椅、凉亭和观光塔的休息平台融入自然系统中，为游客提供学习、娱乐和景观审美体验，并设计了一个环境解说系统以帮助游客理解这些地方的自然和文化含义。场地中最具标志性的建筑物是暖色的彩虹桥，它与当地常见的凉爽湿润天气形成对比。这座长堤连接中心湿地（湖）的三岸，营造了散步及聚会的舒适环境。这里迅速成为备受当地民众和远近游客喜爱的社交和休闲场所。

结论

得益于这些景观技术，衰退的水系统和城市周边的废弃地被成功地转变为高效能、低维护的"城市前厅"。它巧妙地调蓄雨水、净化地表污水、修复原生栖息地，并吸引了广大的居民和游客。2013年明湖湿地公园被官方指定为"中国国家级湿地公园"。

This project is situated in Liupanshui City, along the Shuicheng River. The scope of the task includes ecological restoration of the river, the upgrading of urban open space system, as well as increasing the value of urban waterfront land. The landscape along the Shuicheng River is therefore recovered as an ecological infrastructure providing ecological services to the region. After nearly three years of phase-one design and construction, the once highly polluted and channelized waterway of Liupanshui City has been transformed back into the lifeline of the city through the use of vegetation and natural embankments.

Challenges and Objectives

Liupanshui, known for its cool plateau climate, is an industrial city built in mid 1960s in a valley surrounded by limestone hills, with Shuicheng River running through it. With an area of 60 square kilometers, the city is densely inhabited by a population of 0.6 million. As an element of a major campaign of environmental improvement, the city government commissioned landscape architect to develop a holistic strategy to address multiple serious problems as follows.

(1) Water pollution: As one of the major heavy industrial cities built during the cold war period, Liupanshui has been dominated by coal, steel and cement industries. Consequently, the citizens have suffered with the resulting air and water pollution for a long time. From the industrial chimneys, decades of air pollution deposits fell onto the surrounding slopes and washed into the river along with the storm water that also carries the chemical fertilizer runoffs from the farm land on the slopes and sewage from the scattered settlements on the slope;

场地鸟瞰图及建设前后场地景观对比
Aerial photo of the site, comparisons of before and after photos demonstrate dramatic changes on the site

(2) Flood and storm water inundation: Situated in the valley, the city is subject to floods and storm water inundation during the monsoon season, but also severe drought in the dry season due to the porous limestone geology;

(3) Recovery of the mother river: Channelization of the Shuicheng River was carried out in the 1970s as a solution to inundation and flooding. The channel transmitted the storm water from upstream but caused even more severe flooding problems downstream. Hence, the former meandering mother river became an ugly concrete, lifeless ditch and its capacity for flood retention and environmental remediation was totally lost;

(4) Creation of public space: Recreation and green spaces are inadequate due to the population explosion in the city. The water system that was once a blessing to the city has become a deserted backyard, garbage dump and the dangerous backside of the city. Pedestrian access to a restored green space system is badly needed in such a densely populated community.

The strategy is to slow the flow of water from the hillside slopes and create a water-based ecological infrastructure that will retain and remediate the storm water, and make water an active agent in regenerating a healthy ecosystem to provide natural and cultural services that transform the industrial city into a livable human habitat.

Design Strategies

The submitted Liupanshui Minghu Wetland Park project, 90 hectares (222 acres) in size, is the first phase and a major part of the comprehensive ecological infrastructure project planned for the city by the landscape architect.

For the overall ecological infrastructure, the landscape architect focused both on the Shuicheng River drainage basin and the city. Firstly, existing streams, wetlands, and low-lying land are all integrated into a storm water management and ecological purification system linked by the river, forming a series of water retention ponds and purification wetlands with different capacities. This approach not only minimizes urban flooding but also increases the base flow to sustain river water flow after the rainy season. Secondly, the concrete embankment of the channelized river was removed. A natural riverbank was restored to revitalize the riparian ecology and maximize the river's self-purification capacity. Thirdly, continuous public spaces were created to contain pedestrian and bicycle paths increasing access to the riverfront. These corridors integrate the urban recreation and ecological spaces. Lastly, the project combines waterfront development and river restoration. The ecological infrastructure catalyzes urban renewal efforts in Liupanshui, significantly increases land values, and enhances urban vitality.

As one of the major projects included in the ecological infrastructure of Liupanshui, the Minghu Wetland Park features ecological restoration of the upper stream section of the channelized river. Minghu Wetland Park was created on a site composed of deteriorated wetland patches, abandoned fish ponds and strips of mismanaged corn fields. Its pre-development condition was dominated by garbage dumps and polluted water. As a demonstration of the ecological infrastructure project, this first phase project was designed using all of the tactics for rebuilding ecological health leading to the recovery of biodiversity and native habitat, retention and water quality improvement of storm water, and public access to high quality open space, and finally a catalyst for urban development. The specific park elements that are achieve these objectives are listed below.

夏季，山谷周边细致分级的生态草沟与蓄水池系统起到了"绿色海绵"的效果。雨水被拦截和保留下来，用于捕获或改造农业和城市的非点源污染物。丰富的景观营造了多种多样的栖息地，增加了该地区的生物多样性。

Summer: the inter-locked bio-swales system along the valley acts as "green sponge" that catches the storm water and filtrates the pollutants washed in from the agricultural fields and other non-point pollution sources on the slopes and creates diverse native habitats for biodiversity.

这是水城河上半部分的典型断面图。拆除了之前的混凝土河渠，设计了有繁茂植被的自然渠道，以减缓山上流下来的水速，恢复河岸的生机，成为备受游客喜爱的钓鱼区，而两边的人行道和自行车道也有了其他用途。漂浮的植被能够过滤固体物质以及消除从斜坡上的农田和其他非点源污染区冲刷下来的营养物质。

This is a typical scene of the upper section of the ecologically restored Shuicheng River, featuring the concrete channel been replaced with lushly vegetated stream that slows the storm water flow from the mountains and filtrate nutrients washed in from the agricultural fields and other non-point pollution sources on the slopes, with pedestrian and cycling paths on both sides. A former lifeless river becomes a favorite place for sports fishing.

(1) The concrete river embankment was removed to create two ecological zones. One encourages native vegetation to grow within the flood zone and the other establishes conditions for emergent vegetation in the riverbed. Aerating cascades were created along the river to add oxygen that fosters bio-remediation of the nutrient-rich water.

(2) Terraced wetlands and retention ponds were created to reduce peak water flow and regulate the seasonal rainwater. The terraces are inspired by the local farming techniques that catch and retain water and transform steep slopes into productive fields. Their positions, forms and depths were based on geographic information and a water flow analysis. Native vegetation was planted (mostly sown) to establish associations adapted to the various water and soil conditions. These terraced habitats slow the flow of water and speed nutrient removal from the water by microorganism and plant species that use excess nutrients as resources for rapid growth.

(3) Pedestrian paths and bicycle routes are overlaid on the green spaces along the waterways and form a circuit around and between the wetland terraces. Resting platforms with abundant seats, pavilions and a viewing tower are integrated into the designed natural system for universal access. This fosters learning, recreational and aesthetic landscape experiences. An environmental interpretation system was designed to help visitors understand the natural and cultural meaning of the places. Clearly, the most iconic built artifact is a warm-colored rainbow bridge, in contrast with the frequently cool and damp climate. This causeway connects three sides of the central wetland (lake), creating unforgettable walking and gathering places. These have quickly become favored social and recreational environments of the citizens and attract visitors from near and far.

Conclusion

Through these landscape techniques, the deteriorated water system and peri-urban wasteland has been successfully transformed into a high-performance and low maintenance municipal front yard. It beautifully regulates storm water, cleans contaminated water, restores native habitats for biodiversity, and attracts residents and tourists. It was officially designated as a National Wetland Park in China in 2013.

每个生态洼地的水位都经过精心设计，控制其入水和出水高程。流经水池和植被的水流被控制得很缓慢，以保证净化和去营养过程的进行。堤坝把水池中的水分流、分层，从而减缓水流经公园的速度。设计让每个水池都存蓄足够的水，不仅维持了适宜植被的生长，也创造了引人入胜的环境，四季皆可供游客赏玩。

The water table in each of the bio-swale is carefully designed through controlling the inlets and outlets of water flow and through the dykes that filtrate the water. The slowdown flow makes individual ponds to retain sufficient water to sustain adaptive vegetation and create attractive habitats enjoyed by visitor around seasons.

秋季，应用中的湿地吸引成千上万名游客来此地游玩，不仅有本市的，还有其他地区的。游客和当地人一样，喜欢欣赏秋季富有质感与多彩的景象。多年生的花卉在小路沿线和生态草沟之间，形成了低维护的地表覆盖物，游客在此进行一场场生动、愉快的散步之旅。

Autumn: the working wetland attracts thousands of visitors every day from the city and the far-reaching region. The vibrant colors in autumn have become the "eye candy" to the visitors. The low maintenance vegetation of perennial flowers and grass add more pleasures when walking among them.

道路和生态蓄水洼地之间种植着可自我繁育的花卉，营造了低维护的地表景观。这些元素为行人提供了充满生气、愉悦身心的步行体验。
Self-reproductive flowers are sewed along the paths in between bio-swales, to create a low maintenance ground cover, and create a pleasant walking experience.

夏季，游步道网络穿插在生态蓄水洼地之间，游客可与大自然亲密接触。这是一个将大自然融入其中的体验式环境网络。
Summer: A network of pedestrian paths is weaved into the bio-swales system, allowing visitors to have intimate contact with the living nature. An environmental interpretation system is integrated into this experiential network.

生态洼地的样式和尺寸经过精心设计，与坡地的各种坡度相配合。坡度最陡的地方，生态洼地也最窄。按照最小干预原则，坡地的建造采用随挖随填的方式。

The form and sizes of the bio-swales are adaptive to various degrees of steepness of the terrain, the sleeper the narrower, and are created through cut-and-fill tactics following the minimum intervention principle.

彩虹桥是标志性的文化景观元素，既可俯瞰城市周边广阔的喀斯特地貌景观，也可为游客提供文化之旅，使其感受和体验不同寻常的自然景观。

The rainbow bridge is an iconic cultural landscape element that en-frames the extensive karst landscape surrounding the city, providing a cultural route for experiencing and interpreting the otherwise ordinary natural landscape.

彩虹桥高架在湿地花园上，是前往保护区湿地的通道，也是一个连接通道，让一直繁忙的居民放慢脚步，欣赏城市周围的景象，还有那些过去几十年来被人遗忘和未发现的美景。

A rainbow bridge flies above the wetland park acts an access into the designed wetland, and as a linkage that invite the ever busy residents to "slowdown" their pace to enjoy the everyday landscape surrounding the city, the beauty of which has been forgotten and misused over the past decades.

人们在被一系列生态洼地和阶地环境的清澈的水池边嬉戏，亲密地接触自我繁育的丛丛野花，异常欣喜。

People enjoy the clean water filtrated through the series of bio-swales and terraces, and get excited at the intimate contact with the massive self-reproductive flowers.

044

东营市河口区湖滨新区鸣翠湖景观设计

MINGCUI LAKE LANDSCAPE DESIGN, LAKESIDE NEW ZONE, HEKOU DISTRICT, DONGYING

项目位置：中国山东省东营市
项目规模：97 ha
设计公司：艾奕康环境规划设计（上海）有限公司
翻　　译：张旭

Location: Dongying, Shandong Province, China
Size: 97 ha
Design Firm: AECOM Group
Translated by Zhang xu

该项目"海绵城市"的建设途径主要包括以下三方面。

对城市原有生态系统的保护

最大限度地保护原有的河流、湖泊、湿地等水生态敏感区，留有涵养水源充足、应对较大强度降雨的林地、草地、湖泊、湿地，维持城市开发前的自然水文特征，遵循"海绵城市"建设的基本要求。

生态恢复和修复

对已经受到破坏的水体和其他自然环境，运用生态手段进行恢复和修复，并维持一定比例的生态空间。

低影响开发

根据对城市生态环境影响最低的开发建设原则，合理控制开发强度，在场地中保留足够的生态用地，控制城市不透水面积比例，最大限度地减少对城市原有水生态环境的破坏，同时，根据需求适当开挖河湖沟渠，增加水域面积，促进雨水的积存、渗透和净化。

雨水径流管理根据鸣翠湖主要污染源分析，黄河引水和直接降雨的水质相对较优，而雨水径流水质可能受周边地块开发影响，对鸣翠湖水质造成一定影响。

对周边地块采取最佳雨水径流管理措施，如生态滤水带、雨水花园、绿色屋顶等，收集并处理路面、屋面、停车场以及其他不渗透表面的雨水径流，以改善径流水质，同时生态雨水管理设施有利于雨水下渗补给地下水，对长期改善土壤盐碱化有一定作用。

沿湖带雨水径流管理设计由于用地类型、场地情况的不同，对沿湖雨水径流管理进行分区治理，分别采取相应的生态雨水径流管理措施，以保护鸣翠湖的水质。水质主要分成以下三类。

（1）生态滤水带：大部分临湖区域结合景观设计生态滤水带，截留并处理周边地表雨水径流，经过植物根系以及滤料层净化后排入鸣翠湖，保护鸣翠湖水质的同时为鸣翠湖提供一定量的淡水资源。

（2）生态滤水带+雨水存储设施：部分区域由于场地标高的限制，排水入鸣翠湖存在一定难度，建议设置雨水存储设施，周边地表径流经生态滤水带处理后排雨水罐存储，之后可回用于周边绿化灌溉。

（3）湿地：鸣翠湖西岸用地以生态绿地为主，本身径流水质较优，同时湿地系统的设置为水质净化提供了保证。草桥沟汛期可能同时受强降雨以及风暴潮影响，排洪不畅，产生内涝问题，此时草桥沟水位急剧上涨，部分区域洪水位高达5 m多，鸣翠湖紧邻草桥沟，因此与草桥沟之间的堤岸设置必须满足草桥沟防洪要求。

李陀水库（鸣翠湖）在草桥沟侧约8m的堤岸高程，约8 m左右的堤岸足以抵挡草桥沟洪水，保证鸣翠湖不受草桥沟洪水的影响。鸣翠湖排涝纳入环城水系排涝体系，洪涝时启用强排泵，以确保排涝安全。排水口设计应确保鸣翠湖排水顺畅，可设置溢流坝进行控制，平时控制水位，当鸣翠湖水位超过3 m时，溢流排水。为缓解区域的防洪排涝压力，建议鸣翠湖设计一定的蓄滞洪空间，湖区平均堤岸高程约4.5 m，扣除0.5 m的安全超高后，水位处于3~4 m的空间均可作为滞洪空间。

综上，按50年一遇防洪排涝标准设计溢流坝尺寸：坝高3 m，鸣翠湖水位高于3.0 m时即开始溢流，宽不小于25 m，与下游环城水系相连，以确保鸣翠湖湖体排涝安全。

The three main measures adopted in Sponge City Construction.

Protecting the existing urban ecological system

Maximizing the protection to the original sensitive aquatic zone, as rivers, lakes, wetland, etc., remaining sufficient water conservation area and the resilient forests, grassland, lakes, wetlands which can resist heavy storm. Maintaining the natural hydraulic features before city development, keep tight to the principles of sponge city construction.

Ecological restoration and repairmen

Applying ecological restore & repair method to the water bodies and other natural environment that had been damaged, maintaining a proper proportion of ecological space.

Low impact development

In accordance with the vision of development & construction with low impact on the ecological environment, controlling the development intensity properly, maintaining sufficient ecological land on site, limiting the ratio of the impervious area, minimizing the damage to the urban existing aquatic ecological environment. Meanwhile, excavating channels and lakes in line with reasonable demands, increasing the area of aquatic space, promoting the rainwater treatment process of retention, filtration and purification.

The stormwater runoff management is based on an analysis of the main pollutants' source in Cuiming Lake. The water drawn from Yellow River and the direct precipitation has better quality comparatively. However, been influenced by the development of the surrounding site parcels, the deteriorated runoff water would place a negative impact on water quality of Mingcui Lake.

Carrying on an optimal runoff management measure on the nearby site. For instance, the ecological water treatment belt, rain garden, green roof. Collecting and disposing the runoff from the streets, roofs, parking places and other impervious areas, and purifying the runoff water quality. Meanwhile, the ecological stormwater management facilities along the lakeside could help to recharge ground water and ameliorate the saline soil condition in a long turn.

In the lakeside runoff management design, based on different types and distinct context of each site, the site was divided into different area. In this circumstances, the subarea management was implemented, specific design approach was apply to each area, in order to protect the water quality in Cuiming Lake.Three are there main categories for the water quality,– shown as below.

(1). Ecological water treatment belt: in most of the lakeside area, the landscape design is integrate into the ecological water treatment belt, retaining and disposing the runoff from the surrounding area. And the purified runoff would be discharged into Cuiming Lake after flow through the plant'roots system and the filter material layer. It is not only protecting the water quality in the Cuiming Lake, but also serving as the source of certain amount of fresh water.

(2). Ecological water treatment belt & Storm water storage facilities: Due to the limitation of the site elevation, the runoff from some highland area is difficult to be discharged into Cuiming Lake. So a set of stormwater storage facility is supposed to be placed on the site. The runoff from the surrounding area would be collected in the storm water tank after flew through the water treatment belt, and the stored stormwater could be reused for the surrounding greenery irrigation.

(3). Wetland: the land on the west bank of Cuiming Lake is dominant by the ecological green space. For this reason, the original runoff was relatively

绿地
Greenland

clean, and the construction of the wetland system laid a profound base for the water purification. In the flood season, the straw bridge ditch would be influence by the heavy precipitation and storm tide, resulting in stagnation and waterlogging. In this period, the water level in straw bridge ditch elevated in a rapid rate, the flood water level in some area climaxed to 5 meters high. Cuiming Lake is just adjacent to Straw-Bridge ditch, therefore, the embankment between the Straw-Bridge ditch must satisfy the flood prevent requirement.

It is suggest to reserve the original embankment elevation of Lituo reservoir along the Straw-Bridge ditch, the proximately 8 meters embankment is higher enough to resist the flood from the ditch, making sure Mingcui Lake could be free from the floods'negative influence. Cuiming Lake is an impartible and organic part of the round-city flood drainage system. The potent discharge pump would be used during the flood period to ensure the drainage security. Constructing the overflow weir, controlling the water level in the common day and installing the outlet port to avoid the impeded drainage. When the water level surpassed 3 meters, the storm water would be overflow. In order to relieve the flood drainage stress, the design team prepared some flood retention & detention space the Cuiming Lake. The average water level in Cuiming Lake is proximately 4.5 meters, deduct the reserved 0.5 meters for the security concern, the fluctuated water level ranges from 3 meters to 4 meters, the space between this two elevations could conserve the flood.

In general, the size and dimension in the design is accord with the controlling and drainage standards of the once 50 years rain. The height of the dam is 3 meters high, the lake water would overflow when the water level surpass 3.0 meters, the breadth of overflow paths is no less than 25 meters, and connected with the round city river system in the downstream, ensuring the flood drainage security of Cuiming Lake body.

生态驳岸
Ecological revetment

绿地
Greenland

雨水存储
Rainwater storage

雨水花园
Rain garden

台地储水
Bench terrace water storage

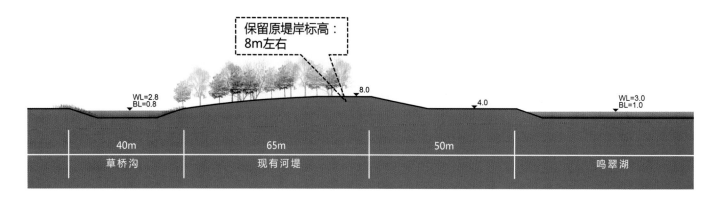

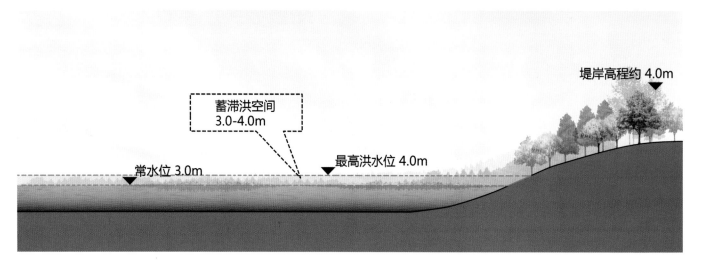

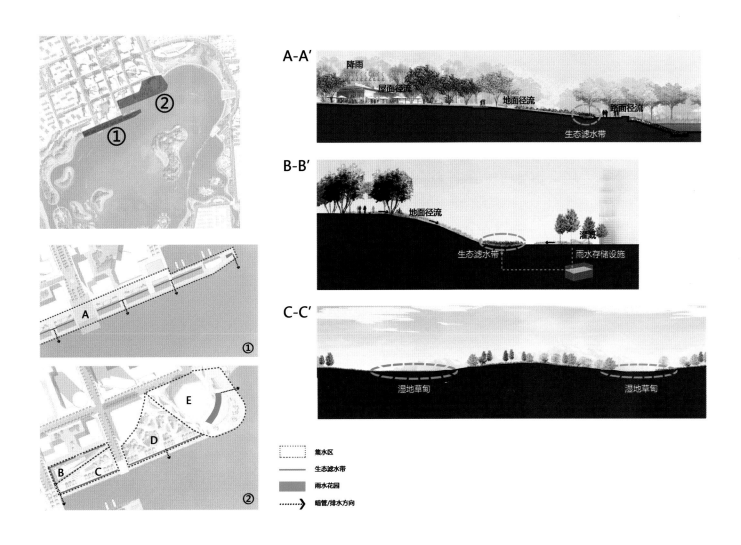

生态滤水带/雨水花园典型剖面

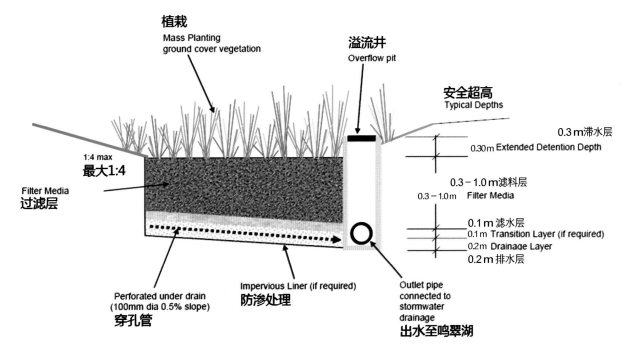

生态滤水带／雨水花园典型剖面
Typical section of ecological filter belt / rain garden

052

哥本哈根城市排水防涝规划

URBAN DRAINAGE & WATERLOGGING PREVENTING PLANNING, COPENHAGEN

项目类型：战略性洪水规划
项目位置：丹麦哥本哈根市
项目规模：3400 ha
设计公司：安博·戴水道
翻　　译：张旭

Project Type: Strategic Flood-Controlling Planning
Location: Copenhagen, Denmark
Size: 3,400 ha
Design Firm: RAMBOLL STUDIO DREISEITL
Translated by Zhang Xu

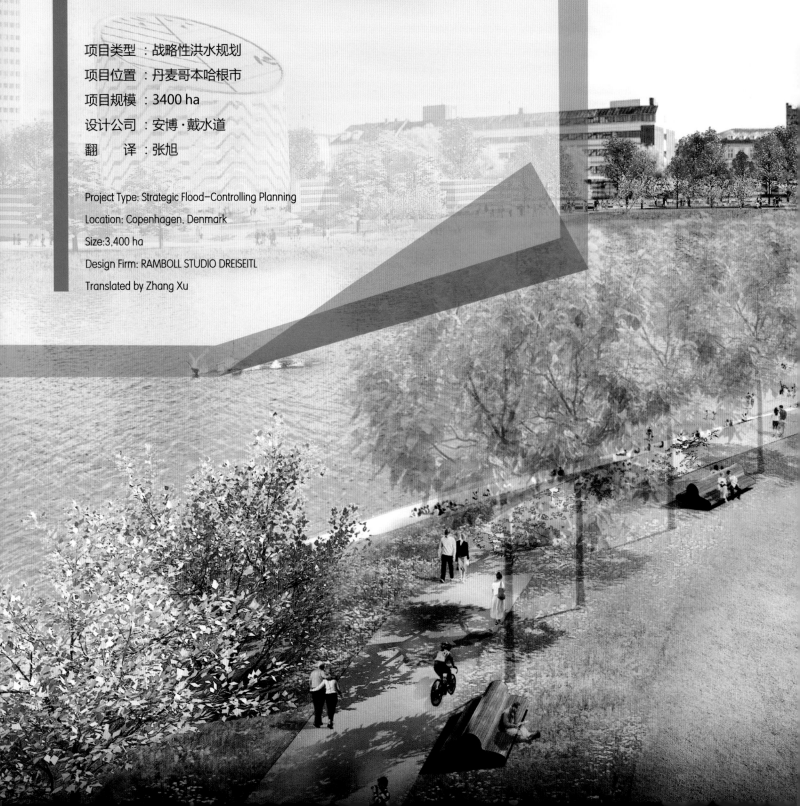

德国戴水道设计公司和安博国际工程集团于2013年共同制订了哥本哈根排水防涝规划,这个案例展现了在面临气候变化的丹麦首都在最新一轮城市规划中如何将"海绵城市"设计与城市排水防涝和水系规划相结合,打造丰富且有效的蓝绿基础设施。该案例同时展示了城市规划在明确街道暴雨管理指标、统筹绿地空间、生物空间和水利空间等方面,如何采用战略可行的方法。该项目在2013年赢得了国际知名的设计奖项INDEX设计奖。

背景

作为全球最可持续的城市的哥本哈根位于气候变化前沿,近些年来遭受了越来越多的暴雨袭击。从2010年8月至2011年8月一年间,哥本哈根市遭受了3次暴雨袭击,主要公路及城市基础设施被淹。2011年7月2日全市大面积区域遭受严重洪涝灾害侵袭,暴雨淹没了城市中心区域大部分的城市街道和地下空间。最初的经济学分析指出,如果不采取任何措施,在未来的一百年间由于气候变化引起天气的剧烈变化造成暴雨事件的破坏力将造成3倍的损失。以此事件作为城市创新设计的契机,哥本哈根决定制订综合的气候适应型策略以保护城市,暴雨管理规划方案便由此而生,它在保护城市有能力抵御未来的暴雨事件的同时,为哥本哈根提供更多的蓝绿空间,增加城市的生物多样性并为市民提供更多的休闲空间。

水文

该暴雨管理规划计划在未来30年间完成,它充分考虑了气候变化可能带来的极端天气情况,以保护哥本哈根市抵御百年一遇的暴雨侵袭。应对百年一遇的暴雨,该规划可容纳城市道路雨水10 cm高度的提升,且计划分担城市排水系统30%~40%的雨水泄流,这样的指标设定是针对气候变化导致的极端暴雨预期增加的40%的降雨量。该暴雨管理规划完全建立在简单的原则之上,要点在于将雨水蓄留于地面之上进行管控,而非增加地下管道升级改造的高昂费用。

哥本哈根市被划分为8块区域,针对各不同集水区进行具象化的规划。安博国际工程集团联合德国戴水道设计公司参与了其中4块区域的规划设计,针对街道、蓄留街道、绿色街区进行了丰富和细致的分析,形象地阐释了规划设计方案如何支撑城市的总体规划目标,为市民增加宜居安全感和舒适度。多功能的空间设计是具象化规划的关键因素,例如,公园和广场在暴雨时可为泄洪场地,而在旱季则可充当市民的休闲娱乐空间。在人口密集和空间稀缺的都市里,这些多功能的空间在99%的时间里供游乐和休闲之用。

海绵手段

哥本哈根城市由两个行政部分组成:哥本哈根和腓特烈斯贝。过去30多年以来,安博国际工程集团一直在规划和设计城市污水处理基础设施系统以及为城市服务的两家废水处理厂。工作的目标即完全控制污水的排放,实现污水的全部处理,营造拥有优良水质条件的港口和海岸环境。也因如此,哥本哈根市已经拥有详细的各区位水利基础设施的数据,并借助综合的计算机建模预测进入两家废水处理厂的雨水及其径流以及雨水溢流至溪流和港口的流量。2011年的暴雨事件也恰巧为该数据模型提供了实际检验机会。

基于已有信息和模型计算,哥本哈根暴雨管理规划得以开展。以腓特烈斯贝行政区为例,在气候适应框架之下进行的暴雨管理规划重点关注城市的低洼部分,确保未来城市开发的可能性,并建设可持续性的排水基础设施,以保证暴雨期间洪水水位10 cm以下。该战略规划基于以下三个原则:

(1)在高地势地段滞留雨水,以保护低洼地段的安全。
(2)在低洼区域采用可靠灵活的雨水径流排放方式。
(3)在此低洼区域进行雨水径流管理。

暴雨总平面图
Rainstorm site plan

晴天鸟瞰图
Aerial view in sunny day

雨天鸟瞰图
Aerial view in rainy day

通过这些战略规划，热点区域被确定下来，一系列暴雨管理规划项目在兼顾考虑其他城市规划目的的前提下也被确定下来，如此暴雨规划之中各项目便与交通基础设施改善规划及城市更新规划相整合。

腓特烈斯贝暴雨管理规划项目包含以下要素：暴雨街道规划、滞留街区规划、中央滞留带规划（公园和广场）、绿道规划、传统暴雨排水管道规划。

腓特烈斯贝林荫道下方的地下通道供暴雨期间的雨水排放之用。作为哥本哈根城市之中景色优美、具有历史意义的林荫路，它的美丽景色以及功能发挥受到车辆停靠的挑战。该林荫道的重新设计包含增大步行道路面积，并将停车区域设置于地下的两条通道之中。在遭遇大雨时，这些通道可作为泄洪场地。在暴雨时节，雨水从中央绿带流向街道，进而随街道坡度变化流向建筑体，对于建筑底部排水管道排水压力的缓解毫无帮助。

安博公司设计的道路模板是改变整体的道路特征，形成V字形道路，在道路中央的绿带中创建大容量的雨水蓄留空间。在下雨之时，雨水从周边的房屋和街道流向该绿色空间，这种在暴雨之时产生的"城市河流"能够容纳每秒3300 m³的雨水量。在常规降雨和干燥季节时，该低洼的绿带同样可作为周边市民休闲娱乐的场所。

中央滞留区域同样是该暴雨管理规划的核心元素。一项全新的建议即通过降低湖内的水位，将哥本哈根三大内城湖泊之一的圣乔治湖转变为海滨公园。在湖岸区域，一处新型的公园将被建造，提供休闲场地，例如散步和慢跑道路，一片可享受晴朗天气日光普照的草坪以及一个嬉水游乐场。新建公园将有效提升生物多样性指数并有效调节环境微气候。汇入湖体的水质将被提升，以改善湖体的生态特性。一处新广场将建于天文馆对面，一个次级街道的绿化带也被包含在内。该公园在暴雨期间被淹没用于泄洪。如此，便形成了收集雨水的庞大空间，并提升了城市的休闲价值。通过这种以休闲娱乐结合雨水排放的方式比通过建造体量巨大、费用昂贵的暴雨排水管道来转移50万 m³预期的雨水的方式节约近1.34亿欧元用于建造地下雨水管道的费用。

效益

哥本哈根暴雨管理规划以及同系列的项目设计计划的通过首先经由严格的审查和讨论过程，然后进入优先考虑程序以及决策阶段。正如哥本哈根的其他规划和建设项目，该过程包含所有相关的市政利益相关者的参与，并允许公众意见的反映。安博公司为此决策审核通过的过程进行了详尽的成本效益评估。这个评估权衡了在例如资金以及新设施的运营之中投入与收益的关系：它能够保护城市抵御洪水侵袭，增加的绿地空间可降低空气污染，降低对于现存下水系统改造的费用投入，提升房产价值。2015年有几百个具体项目获批，招投标已经开始进行。

社会经济学分析结果显示，这种关注提升城市整体宜居性的设计所产生的费用大大超越了其建造和维护的费用。尽管在30年的时间中需要投资13亿欧元的费用，它在预防城市洪涝灾害和降低损害方面的收益将大大超越其投资。这一案例展示了长效、全面、可持续的城市问题解决模式，并可作为世界上其他同样面对此类问题城市的规划范本。哥本哈根市采用此种方式抵御气候变化，能够得到社会经济方面的诸多益处。

哥本哈根暴雨管理规划展示了在一个多学科团队之中各传统专业跨专业合作的需求。水利工程师与建模专家合作，以便管理复杂的水利以及工程水利系统之中的技术局限。景观设计和规划师在项目之中提供全新的蓝绿城市环境设计，而经济学家则提供决策过程之中所需要的成本效益评估。哥本哈根暴雨管理规划能够提升城市的可持续性，并为城市提升其环境质量、保障居民生活品质、确保城市长期的弹性适应能力和经济增长趋势增添重要的元素。

雨天景观
Landscape in rain

BEFORE: Barriers

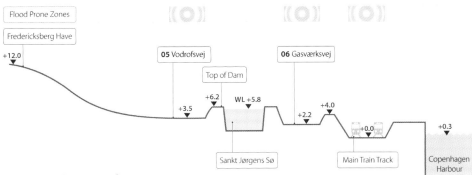

AFTER: Connections

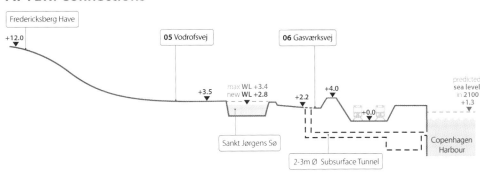

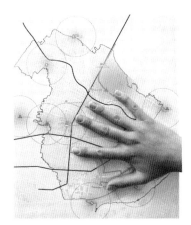

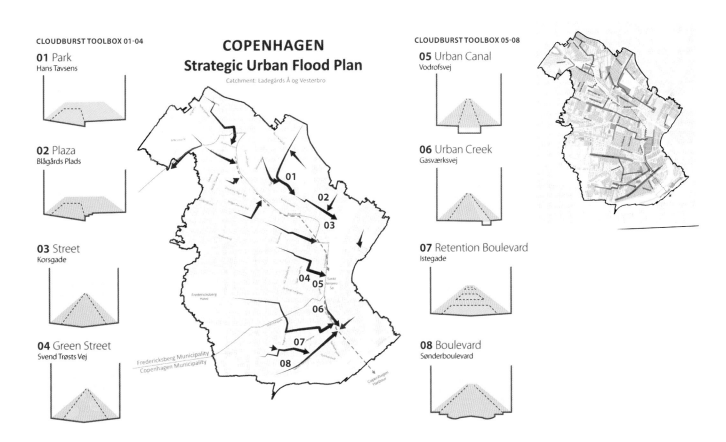

CLOUDBURST TOOLBOX 01-04

01 Park
Hans Tavsens

02 Plaza
Blågårds Plads

03 Street
Korsgade

04 Green Street
Svend Trøsts Vej

COPENHAGEN
Strategic Urban Flood Plan
Catchment: Ladegårds Å og Vesterbro

CLOUDBURST TOOLBOX 05-08

05 Urban Canal
Vodrofsvej

06 Urban Creek
Gasværksvej

07 Retention Boulevard
Istegade

08 Boulevard
Sønderboulevard

Studio Dreiseitl and RAMBOLL International Engineering Group collaborated in compiling the urban water drainage & waterlogging preventing plan. This case demonstrates how to combine the sponge urban design with urban drainage & waterlogging preventing planning in the latest round of Danish urban planning, and how to how to construct extensive and effective blue and green infrastructure in the face of climate change. This case illustrates how to adopt strategic and feasible approach following the cleared streets stormwater management indicators on the balanced green space, biological space and hydraulic space integration. The project has won the internationally renowned INDEX Award in 2013.

Background

As the world's most sustainable city, Copenhagen is in the forefront in terms of the climate change. In recent years the city has experienced a growing number of storm events. From the August of 2010 to the August of 2011, Copenhagen was attacked by three heavy storm, major highways and urban infrastructure were submerged. A large area of the city severely suffered from the flooding disasters on July 2nd, 2011. Storm flooded most of the city streets and underground space in the central urban area. The initial economic analysis pointed out that if no measures ware taken, the devastating storm result from dramatic climate change would cause three times the damage in the next one hundred years. This event has been perceived as an opportunity for an innovative urban design, Copenhagen decided to develop comprehensive climate-adaptive strategies to protect the city. The stormwater management plan was produced in this context. It has the ability to prevent the city from future damage caused by storm, and it is capable to offer more blue-green space. Meanwhile, it could increase urban biodiversity and provide more recreational space for citizens as well.

Hydrology

The plan of stormwater management is expected to be accomplished in the next 30 years, which fully takes the extreme weather conditions caused by climate change into account, it can protect Copenhagen against the heavy rains of once a year. In order to be efficiently responsive to the once a hundred years'rain, the plan could accommodate 10 cm lifted rainwater level on the streets, and it supposed to share and dispose 30%~40% volume of the rainwater discharged by the urban drainage system. The determination of the project'indexes base on the extreme storm precaution, in which is forewarned of the increased 40% amount of precipitation cause by the climate change. The stormwater management plan is entirely on the basis of a simple principle, the main point is that the controlled rainwater should be retained or detained on site, rather than increasing the already highly cost to upgrade or reconstruct the underground pipeline networks.

Copenhagen is divided into eight regions. For every water catchaent areas, a figurative design was delivered. Ramboll International Engineering Group and Atelier Dreiseitl cooperated on the planning & design for four regions. They comprehensively and carefully analyzed the roads, retention streets and green neighborhoods, and illustrates how could the detailed plan & design be supportive to purpose of the master plan vividly, and enhanced the sense of comfort and security for the citizens.Multipurpose space design is a key factor to these figurative planning, such as parks and plazas could be used as the flood discharge area during heavy rains and remained the public recreational space in the dry season. In the densely populated and space deficient city, these versatile spaces will be used for entertainment and leisure purposes in 99% of the time.

Sponge measures

Copenhagen city is composed of two administrative parts: Copenhagen and Frederiksberg. Over the past 30 years, Ramboll Engineering Consultant Firm has been planning and designing the urban sewage treatment infrastructure systems and two wastewater-treatment plants which serve the city. Objectives of the systems are to control the sewage water discharge and to realize the full treatment of sewage water, and to create the ports with excellent water quality and pleasant coastal environment. It is also the reason why Copenhagen already have the detailed data of all the water infrastructure in each region, and using the synthetic computer modeling to predict the volume of the rain and runoff which flow into the two treatment plants and the volume of the overflow which discharged into the streams and the ports. And the 2011 storm event also happened to provide a real opportunity for conducting a test for the statistics model.

The stormwater management plan in Copenhagen can successfully be undertaken mainly because of the available information and the modeling calculations. In Frederiksberg Administrative Region, for example, implementing the stormwater management under the framework of the climate adaptation, the plan should focus on low-lying parts of the city, the remained headspace to reserve and conserve the possibility of future urban development, and the construction of sustainable drainage infrastructure to ensure the flood elevation is beneath 10cm in the circumstance of heavy storm. The strategic planning based on three principles,shown as follows.

(1) Retain rainwater in the high-lying area, to protect the security of the low-lying areas.

(2) Establish a reliable and flexible stormwater runoff discharge system in the low-lying area.

(3) Implement stormwater runoff management in the sub low-lying areas.

Through these strategic planning, the sensitive areas were identified, a series of stormwater management planning projects were determined under a balanced consideration of other specialized planning as a prerequisite, so as an example, the improvement of transportation infrastructure and urban renewal plan were integrated in the stormwater management planning.

Frederiksberg stormwater management planning project consists of the following elements: stormwater management street planning, stormwater retention block planning, central retention zone planning (parks and squares), greenway planning, traditional stormwater drainage planning.

Frederiksberg Avenue - stormwater management street design & planning. The underpass beneath Frederiksberg Avenue was designed as a discharge channel of rainwater during heavy rains. As one of the most scenic, historic avenue in Copenhagen, the vehicle parking challenged its beautiful scenery and function exerting. In the re-design of Frederiksberg Avenue, the greenway's walking paths area will be enlarged, the parking lot will be set in the two underground channels. When the rains are heavy, these channels could be used as flood venue. In wet season, rainwater will flow trough the central green belt to the street, and delivered to the building following the slope variation on the streets, and played little role in relieving the drainage pressure in the bottom of the building.

The modular street model designed by Ramboll Group has changed the overall features and characteristics of the roads, forming a V-shaped path, creating large volume of rainwater retention & detention space in the middle of the green belt on the roads. When it rains, the rainwater from the surrounding roofs and streets can flow into these green spaces. The URBAN RIVER resulted from heavy rains can accommodate and contain 3300 cubic meters per second amount of rainwater. In the condition of regular rain and dry season, the low-lying greenbelt can also be used as entertainment places for the surrounding citizens.

Central retention region is also a core element for the stormwater management planning. A new proposal is to transform St. George Lake into a waterfront park by lowering the water level in the lake. St. George Lake is one of the three great inner city lakes in Copenhagen. In the lakeshore area, a new park will be designed and built to provide leisure venues, such as walking, jogging path. One can enjoy the sunny weather on sun-drenched lawn. A paddling playground is also available. The new park will effectively enhance the index of biodiversity and can be efficiently regulated microclimate in this environment. The quality of the water imported to the lakes will be improved, in expect to promote the ecological characteristics of lakes. And a new plaza will be built opposite to planetarium, the green space design on a secondary street are also included. The park will be submerged during heavy rains play a role in the function of flood drainage. So, this formed a large space to collect rainwater, and also improved the recreational value of the city. In the anticipation of transfer 500,000 m^3 rainwater by creating the large scaled and highly expensed stormwater drainage network combined with the entertainment and recreational function, in this approach it could saved nearly 134 million Euros of reconstructing underground municipal sewer system instead.

Effectiveness

Copenhagen stormwater management planning and a series of related project design planning will first go through the rigorous process of review and discussion, and then enter the program-prioritizing and decision-making stage. Exactly like other planning & construction projects in Copenhagen, the process will involve in all the relevant municipal stakeholders to participate in and allow public opinions to be reflected. Ramboll Engineering Consulting Firm has prepared a detailed cost-benefit assessment for the decision approval during the review process. For instance, this assessment weighed in the budget of income & expenses from the financial investment and new facilities operation. It can protect the city against flooding, alleviate air pollution due to the increase of green space, and reduce the cost of the existing sewer system renovation in return to improve the properties'value. Hundreds of specific projects were approved in 2015. Bid and tender work has begun afterwards.

Social economic analysis shows that the benefits and profits from this kind of project design, which dedicated in promotion of the urban livability, are far beyond the cost of its construction and maintenance. Despite the need of injecting totally 1.3 billion Euros into the project in the following 30 years, the cost of investment could barely compare to the effectiveness yield from the urban flood & waterlogging preventing and the damage & lost minimizing. This case shows a long-term, efficient, comprehensive and sustainable solutions to urban issues, and can serve as a model in dealing with similar problems in other area worldwide. Just like Copenhagen combat climate change in this way, and obtain many benefits in social and economic aspects.

Copenhagen stormwater management plan demonstrated the necessity of multi-disciplinary collaboration among a traditional professional team. The hydraulic engineers cooperated with the latest modeling experts to break through the bottleneck constraints in facing with the management of the complex hydro engineering and hydraulic techniques. Landscape architects and urban planners put forward a brand new blue & green urban environment design for the projects in joint efforts, and economists provide the cost-benefit assessment required by the evaluation and decision making process. Stormwater management planning in Copenhagen can enhance the sustainability of the city and promote the quality of its environment, safeguard the citizens'living standards, ensure long-term flexibility of the city's adaptive capacity, and stimulate and fuel economic growth of the city.

060

绿色海绵般的水适应城市：哈尔滨群力雨洪公园

A GREEN SPONGE FOR A WATER-RESILIENT CITY: QUNLI STORMWATER PARK, HARBIN

项目位置：中国黑龙江省哈尔滨市

项目规模：约30 ha

设计公司：土人景观

获奖信息：2012美国景观设计师协会杰出设计奖

2013世界建筑节-景观项目奖优秀奖

2014 Zumtobel Group奖城市发展与创举类提名奖

2014 罗莎芭芭拉景观奖入围作品

Location: Harbin City, Heilongjiang Province, China

Size: About 30 ha

Design Firm: Turenscape

Awards: 2012 ASLA Award of Excellence in General Design

2013 WAF – Landscape Prize Excellence

2014 Zumtobel Group Award – Urban Development & Initiatives Category Nominated

2014 Rosa Barba International Landscape Prize Finalist

现在的城市并不是"水适应"的,并且地表水的泛滥造成了严重的水涝问题,正确运用景观设计学的原理可有效地治理水涝。雨洪公园可被连接并整合到不同尺度的生态基础设施中,作为绿色海绵来净化和储存城市雨水。

挑战与目标

由于世界各地不断扩大的城市化,且研究表明气候变化导致了前所未有的降雨量增加,因此暴雨导致的城市洪水已经成为全球性问题。在中国,多数城市都处在季风气候中,70%~80%的年降水都集中在夏季,在一些极端的例子中,每年20%的自然降水可以在一天内完成。以北京为例,年平均降水只有500 mm,但在2011年,仅一天的降水量就达到了50~120 mm。因为不渗水铺装的增加,即使在常态降雨情况下,城市雨涝在中国的各主要城市中仍然屡见不鲜。

营造具有海绵作用的景观是常规市政工程以外能对城市雨洪水管理发挥很大作用的优良途径。这种方法的一个例子是土人景观设计的哈尔滨群力雨洪公园,其综合了大尺度雨洪景观管理和乡土生境的保护、填充地下水、居民休憩和审美体验等多种功能,这些是支撑城市可持续发展所必需的生态系统服务。

2006年,哈尔滨市东部新城的群力城区内开始建设,总占地2733 ha。在接下来的13~15年里,将有3200万m^2的建筑全部建成,约30万人将在这里居住。仅有16.4%的城市土地被规划为永久的绿色空间,原先大部分的平坦地将被混凝土覆盖。当地的年降水量是567 mm,60%~70%集中在6-8月份,历史上该地区洪涝频繁。

2009年,受当地政府委托,土人景观承担了群力雨洪公园的规划设计,公园占地34.2 ha,原为一块被保护的区域湿地。受周边道路建设和高密度城市发展的影响,湿地面临着严重威胁。最初委托方只要求设计师能想办法维护湿地的存在,土人的设计改变了为保护而保护的单一目标,而是从解决城市问题出发,利用城市雨洪,将公园转化为城市雨洪公园,从而为城市提供了多重生态系统服务:它可以收集、净化和储存雨水,经湿地净化后的雨水补充地下水含水层。得益于生态和生物条件的改进,该雨洪公园不仅成为城市中一个很受欢迎的游戏绿地,还从省级湿地公园晋升为国家级城市湿地。

设计方案

该项目中,创新性地运用了许多设计战略,具体包括以下几个方面。

(1)保留现存湿地中部的大部分区域,作为自然演替区。

(2)沿四周通过挖填方的平衡技术,打造一系列深浅不一的水坑和高低不一的土丘,成为一条蓝-绿宝石项链,作为核心湿地雨水过滤和净化的缓冲区,形成自然与城市之间的一层过滤膜和体验界面。沿湿地四周布置雨水进水管,收集新城市区的雨水,使其经水泡系统,沉淀和过滤后进入核心区的自然湿地。不同深度的水泡为乡土水生和湿生植物群落提供多样的栖息地,开启自然演替进程。高低不同的土丘上密植白桦林,步道网络穿梭于丘陵和水泡之间,给游客带来穿越林地的体验。水泡中设临水平台和座椅,使游客更加亲近自然。

(3)高架栈桥连接山丘,给游客带来凌驾于树冠之上的体验。多个观光平台,5个亭子(竹、木、砖、石和金属)和两个观光塔(一个是钢质高塔,

平面图
Site plan

位于东部角落里；另外一个是木质的树状高塔，坐落在西北角）。在山丘之上，由空中走廊连接，通过这些体验空间的设计，游客远可眺公园之泱泱美景，近可体验公园内各种自然景观之元素。

结论

通过场地的转换设计，湿地的多种功能得以彰显：包括收集、净化、储存雨水和补给地下水。昔日的湿地得到了恢复和改善，乡土生物多样性得以保存，同时为城市居民营造了舒适的居住环境。正因其对生态的显著改善，该公园已晋升成国家级城市湿地公园。

该项目成功地构建了一个雨洪管理样本以及一套可复制可参考的技术，不仅可在全中国运用，在全球面临相似社会与文化问题的国家也可实践。其可推广性在于该套技术具备以下几个特点。

（1）低技术要求与可持续性：该项目展示了城市雨洪管理运用简单技术即可实现，并因使用场地材料易于修建，同时维护上实现可持续性的运作也不再是难题。

（2）节省成本：该雨洪管理生态项目在建造和维护上均成本低廉，因此值得被复制运用，尤其是在发展中国家。当然，发达国家也同样适用。

（3）高效能：不同程度的城市内涝问题，均可采用该雨洪管理模式。该项目充分说明，如果一座城市将10%的土地转变成"绿色海绵"用于吸收雨水，那么它几乎可以解决当代城市中普遍存在的雨涝问题。

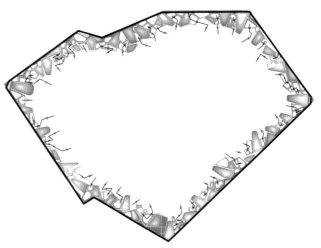

雨水的流向及通过公园周围生态泡的过滤过程
The filtration of storm water through the ponds in the periphery of the park

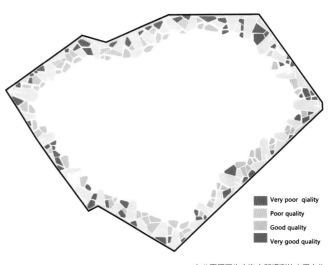

在公园周围生态泡内所观测的水质变化
The improved water quality observed in the filtering ponds

Contemporary cities are not resilient and inundation of surface water poses a substantial problem for them. Landscape architecture can play a key role in addressing this problem. Stormwater parks, connected and integrated into an ecological infrastructure across scales, can act as a green sponge, cleansing and storing urban storm water.

Challenges and Objective

Due to China's ever expanding urbanization, and arguably, climate change leading to unpredictable precipitation, urban flooding caused by storm water has become a global issue. Particularly in China, where most cities have monsoon climate, 70%~80% of the annual precipitation falls in the summer, and in some extreme cases, 20% of the annual rainfall can happen in one single day. Beijing for example, has an average annual precipitation of only about 500 mm (20 inches), but it received 50~120 mm in rainfall in just one day in 2011. Serious urban floods have been hitting the major cities in China even in times of normal rainfall, mainly because of an increase in impermeable paved surfaces.

Using the landscape as a sponge is a good alternative solution for urban storm water management. An example of this approach is demonstrated in Turenscape's storm water park in Harbin, which integrates large-scale urban storm water management with the protection of native habitats, aquifer recharge, recreational use, and aesthetic experience, in all these ways fostering urban development.

Beginning in 2006, a 2,733 hectare new urban district, Qunli New Town, was planned for the eastern outskirts of Harbin in Northern China. 32 million square meters of building floor area will be constructed in the next 13 to 15 years. More than one third of a million people are expected to live there. While about 16.4% of the developable land was zoned as permeable green space, the majority of the former flat plain will be covered with impermeable concrete. The annual rainfall there is 567 millimeters, with the months of June, July and August accounting for 60%~70% of annual precipitation. Floods and waterlogging have occurred frequently in the past.

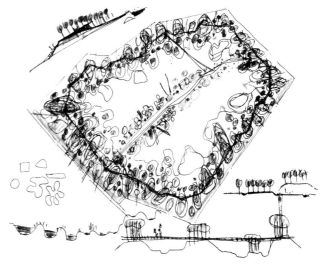

设计概念手绘图
The hand-drawing of design concept

In 2009, Turenscape was commissioned to design a park of 34.2 hectares right in the middle of this new town, which is listed as a protected regional wetland. The site is surrounded on four sides by roads and dense development. This wetland had thereby been severed from its water sources and was under threat. Going beyond the original task of preserving this wetland, Turenscape reconnected water networks and transformed the area into an urban storm water park that will provide multiple ecosystems services. It collects, cleanses, and stores storm water and infiltrates it into the aquifer, and will protect and recover the native habitats, provide a public space for recreational use and aesthetics experience, as well as foster urban development. The stormwater park has not only become a popular urban amenity but also been upgraded to a protected national urban wetland park because of its improvement to ecological and biological conditions.

Design Strategies

Several design strategies and elements have been employed.

(1) The central part of the existing wetland is left along to allow the natural habitats to continue to evolve.

(2) Earth is excavated and used to build up an outer ring-a necklace of ponds and mounds. The ring acts as a storm water filtrating and cleansing buffer zone for the core wetland, and a transition between nature and city. Storm water from the newly built urban area is collected into a pipe around the perimeter of the wetland and then released evenly into the wetland after having being filtered through the ponds. Native wetland grasses and meadows are grown next to ponds of various depths, and natural processes are initiated. Groves of native silver Birch trees (Betula pendula) grow on mounds of various heights and create a dense woodland. A network of paths links the ring of ponds and mounds, allowing visitors to have a walking-through-forest experience. Platforms and seats are put near the ponds to enable people to have close contact with nature.

(3) A skywalk links the scattered mounds allowing visitors to have an above-the-wetland experience. Platforms, five pavilions (bamboo, wood, brick, stone, and metal), and two viewing towers (one made of steel and located at the east corner, the other made of wood and looking like a tree at the northwest corner) are set on the mounds and connected by the skywalk, allowing views into the distance and observation of nature in the center of the park.

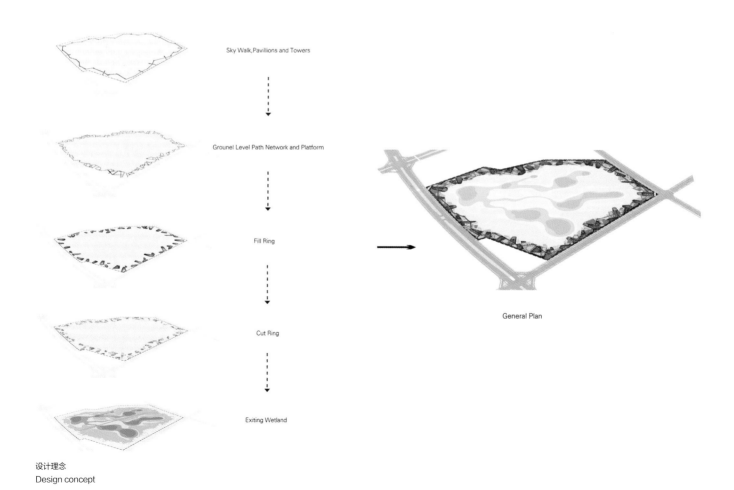

设计理念
Design concept

雨洪公园鸟瞰图（夏季，面向西）。雨洪公园促进了周边地产的开发。
This is the aerial view (toward the west) of the storm water park in the summer. The park has catalyzed the residential development around it.

公园东部边缘挖出的雨水过滤池塘
The filtering ponds on the eastern edge of the park created by excavation

公园外围作为城市与自然之间的过滤带（面向东）
The periphery of the park acts as a border between city and nature
（view toward the east）

面向西的景观
View toward the west

空中走廊
The skywalk

公园北部边缘与雨水过滤池整合在一起，形成小路和平台，使参观者融入湿地并体验自然。
Paths and platforms integrated with filtering ponds on the northern edge allow visitors to penetrate the wetland and experience nature.

架设于西边过滤池塘之上的空中走廊,通向树状观光塔,塔上可俯瞰整个公园。
The skywalk above the western filtering ponds leading to the tree-like viewing tower, where the visitors can overlook the entire park.

Conclusion

The completely transformed site performs many functions, including collecting, cleansing, and storing storm water, and recharging underground aquifers. The pre-existing wetland habitat has been restored and native biodiversity preserved. Potentially flooding storm water now contributes to an environmental amenity in the city. And the storm water park has been upgraded to a National Urban Wetland Park because of its improvements to ecological and biological conditions.

This stormwater management project has been successful in developing a model and a series of techniques that are replicable and transferable throughout China and can be replicated worldwide, especially in those countries with similar social and cultural conditions due to the following qualities.

(1) Low-tech and sustainable: This demonstrative project showcase the low-tech approach to urban storm water management and is easy to build/implement using local materials and easy to maintain for sustainable operation.

(2) Inexpensive: This ecological model of storm water management is inexpensive to build and maintain, and therefore easy to replicate and hence suitable to be implemented in the developing countries, as well as in the developed countries.

(3) Efficient: This storm water management model is suitable in solving the urban storm water inundation at various scales. This project demonstrates that if a city can allocate 10% of total area as green sponge area for storm-water management, it can virtually solve the storm-water problem that is commonly seen in the contemporary cities.

公园西边由挖掘出泥土堆成小山丘，上植白桦林，形成具有东北地域特色的山林景观，人行其中，有置身山林的感觉。
The mounds with the dirt from the excavations on the park's western periphery create a valley experience and remind people of the regional natural landscape of rolling hills.

木亭,唤起对东北乡土建筑的回忆。
The Wood Pavilion is full of a fade reminiscent of the regional vernacular buildings in Northeast China.

070

自然与文化的重生：在新城中重塑三角洲区域

RE-BORN of NATURE AND CULTURE: RE-ESTABLISHING THE DELTA IN A NEW CITY

项目类型：分析与规划
项目位置：中国广东省佛山市顺德区
项目规模：7200 ha
设计公司：SWA集团
翻　　译：张旭

Project Type: Analysis and Planning
Location: Shunde District, Foshan, Guangdong Province, China
Size: 7,200 ha
Design Firm: SWA Group
Translated by Zhang Xu

SWA Group • Shunde New City

项目陈述

文化：傍水而生

珠江三角洲河域系统的渠化与加固破坏了当地自然景观与文化景观。纵观整个历史进程，顺德是广为人知的傍水而生、依水而栖，甚至家长里短、生活琐事上都有河水流过的痕迹。船只在运河中通航，食物由运河里获取，就连趣闻轶事、流言蜚语都通过运河传播开来。村庄中的无花果树枝叶繁茂，人们在树影下聚会交流，丧失运河意味着丧失更多。因此，恢复自然系统与文化体系成为顺德新城设计的核心任务。

SWA集团提出了不同于任何其他竞赛作品的方案，进一步应对了挑战。方案构思的主题是三角洲生态系统的回归，建设72 km²的人工湿地并将其作为多模态城市的枢纽点，为更大的珠江三角洲区域提供鸟类与野生动物的栖息地，同时增强蓄洪能力并重拾水生文化。在规划中，将单个岛屿开发成适宜人行尺度的多功能村庄，中间由绿带、水道、轨道、路径等环保基础设施与多层次的交通体系相连接。基础工作的提升将促进高端人才的引进和科技驱动的雇佣，这些均可促进当地经济的复兴。

环境：自然建筑

在生态建设的过程中，将边缘表面区域最大化，可以增加营养交换。例如，在健康的湿地中，边缘效应可以增加氧气的运输。然而，在中国，三角洲的渠化固化摧毁了边缘效应表面作用的机制条件，因此降低了河流的健康状态与栖息潜力。

现有的机遇是在生态与经济层面，充分利用紧缺的河岸边缘与水体之间的相互关系。因此，方案所提议的纵横交错的水网与精心设计的水道，在增强过滤能力的同时将新规划的顺德新城与运河系统联系起来。绿带与水道齐头并进，衔接融合湿地，并在其边缘空间中将综合生物沟体系纳入其中。这些生物沟来渗入城市的地表径流。绿带的边缘矗立着体量充盈的新型城市森林，不仅为人们提供了休憩场所，同样也是完成碳吸附作用的重要场所。随着水体边缘空间环境的改善，对于市民来说，滨水的散步、工作、休闲机会都会有所增加。在开放空间系统中规划了相当数量与种类的便利设施，包括城市步道和广场、湿地公园、教育展示中心、运动公园和城市森林。在运河系统中嵌入了住宅区与办公区，增加了河流的生态效益，周边的地价也因此升高。提升的土地价值又反向吸引了创意产业和高素质劳动力。这些水道的充足容量保障了洪水蓄贮能力，增强了栖息地的安全与保障。

经济：城市引擎

顺德当下的经济创收依然依赖于劳动力密集和污染力严重的制造业。随着第二产业逐步脱离中国南方，顺德现有的机遇便是发展有关教育与环保的本土新经济。遵循类似的途径，加州尔湾和加利福尼亚依托高质景观、开放空间系统和高校研究院建立了自身基金会，并最终以此吸引了高新技术人才。以此为鉴，顺德将现有高校作为新兴研究与进取发展的基地进行了

场地规划

该72平方千米的新城规划将湿地三角洲系统的建设融入多模式城市的发展中，为更大尺度的珠江三角洲提供鸟类和野生动物的栖息地，提升区域蓄洪能力。文化、环境、经济都得到了重生。

NATURE AND CULTURE RE-BORN
re-establishing the delta in a new city

1. North Transit Station
2. Preserved Canal Town
3. North Harbor
4. Shunde Old Town
5. Mixed-use Residential
6. Water Village Resort
7. Employment Center
8. Wildlife Preserve
9. Mountain Park
10. South Transit Station
11. Island Residential
12. University Center
13. Financial Center
14. Civic & Culture Center Center
15. Wetland Park

Site Plan
This 72-square-km new-city design weaves a constructed wetland delta system into a multimodal city, restoring bird and wildlife habitat for the larger Pearl River Delta, expanding flood storage capacity, and returning a way of life. Culture, environment, and economy are re-born.

扩张,与此同时,三角洲的重建与开放空间的组合大大吸引了高端人才集聚的产业。城市引擎位于两处以便利交通为导向的多模态中心上。

其他的城市中心配备了存留岛屿、自给自足的居住单元、零售、办公、教育和民用设施。市民文化中心、财务中心、办公园区、学院校园和度假胜地具有明确的区划和显著的功能,且其建设遵循人本主义设计原则。不像其他中国现行的400 m²规划街区,该设计提倡100 m的小型系统。由小街区和小街道组成的邻里空间营造了尺度宜人和适宜步行的环境。建筑就建造在街道边缘,以缩短甚至消除建筑后退红线的方式营造了亲切舒适的空间,这些空间以运河为骨架并被树荫覆盖。

一体化系统

顺德三位一体的骨架中包括水系统、交通系统和湿地公园体系。在顺德新城的很多区域中,水是给予了它们身份的认同。在水上的士系统中,水不仅承载了岛屿间交通往来的通勤功能,同样为娱乐休闲提供了全新模式。水上的士、公交和轨道交通系统中大约每500 m设立一个换乘站点,该系统也同样支持岛屿村落的步行体系。单轨交通环线连接起村庄中心,自行车系统依傍河流、运河和林荫大道走势而建,快速轨道和摆渡提供了广州至香港区域性交通连接。

Project Narrative

CULTURE: Living on the Water

The channeling and consolidation of river systems throughout the Pearl River Delta has destroyed both nature and local culture. Throughout its history, the Shunde region has been widely known as a water-based society, one where daily life revolved around the water's edge. Transportation, food, and daily gossip originated along the canals. Families congregated and socialized under the cool shade and broad welcoming branches of the village ficus trees. The loss of these canals led to the loss of much more. Thus, the restoration of both natural and cultural systems has become the central objective of Shunde's new city design.

We took this challenge one step further by creating a proposal unlike any other in the competition. The idea is to put the delta back; restoring 72 square kilometres of constructed wetlands as the armature for a multimodal city, creating bird and wildlife habitat for the larger Pearl River Delta, while simultaneously expanding flood storage capacity and a lost water-based culture. The plan develops individual islands as pedestrian scaled mixed-use villages that are linked by a proposed environmental infrastructure of greenbelts, water corridors, rail, trials, and a multilayered transportation system. In so doing, local economic conditions are revitalized as the groundwork is set for attracting a highly-educated workforce and technology-driven employment.

environment - architecture of nature

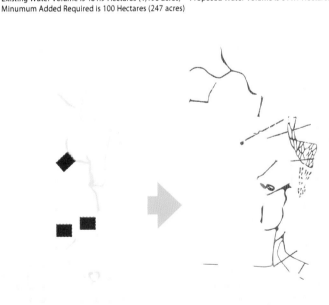

Original Proposed Plan
channelize + drain
Existing Water Volume is 481.9 Hectares (1,190 acres)
Minumum Added Required is 100 Hectares (247 acres)

Regenerative Solution
restore the delta
Proposed Water Volume is 611.1 Hectares (1,510 acres)

Proposed Water System

■ Proposed Waterway / Reservoir
Minumum Added Required is 100 Hectares (247 acres) For Local Flood Control
Existing Waterways

The Chinese government proposed a typical solution to flooding. Creating 1 square kilometer reservoir for local flood control was all that was required.

Alternatively, a variety of waterways are proposed; wide waterways with opportunities for use as major recreation corridors. The waterways provide high volume flood storage capacity, increasing protection and safety for inhabitants.

环境——自然建筑
Environment---Architecture of nature

ENVIRONMENT : Architecture of Nature

In ecological processes, maximizing edge surface area increases the opportunity for the exchange of nutrients; for example, in healthy wetlands, edges increase the transfer of oxygen. However, in China, the channelization of the delta has reduced edge surface conditions; in turn reducing the health and habitat potential of its rivers.

The opportunity exists to exploit these correlations between edge intensity and water-not only ecologically, but economically as well. Thus, a braided system of fine-grained waterways is proposed to increase filtering capacity and reconnect the new planned city of Shunde to the river network. Green belts run parallel to the waterways and incorporate wetlands and a comprehensive bioswale system along its edges. The bioswales serve to filter urban water runoff. The Greenbelt edges are planted with a dense new urban forest, providing both a place of refuge and carbon sequestration. Through the increased water edge condition, more opportunities are developed for the citizens of Shunde to walk, work, and recreate along the river edges. A wide variety of recreational amenities are planned for the open space system, including urban promenades and plazas, wetland parks, educational interpretive centers, sports parks and urban forests. Real estate values increase by weaving residential and office development into the same braided system that benefits the river ecologically. Improved land values in turn attract creative industry and an educated workforce. The waterways provide high volume flood storage capacity, increasing protection and safety for inhabitants.

ECONOMY: City Engine

Shunde's current economy is based upon labor intensive, high-polluting manufacturers. As industry departs Southern China, the opportunity exists for Shunde to grow a new local economy; one based on education and environment. Utilizing a similar approach, Irvine, California built its foundation upon high-quality landscapes, open space systems, and a university that ultimately attracted high-tech employers. Likewise, an existing college in Shunde is expanded as the a basis for new research and development initiatives, while the reconstruction of the delta and resultant open space will attract industry based on a highly-educated workforce. The city engine is anchored by two major transit-oriented multi-modal centers.

Other urban centers comprise the remaining islands, each a self-contained unit of residential, retail, office, educational, or civic uses. A civic and cultural center, financial center, office campus, academic campus, and resort provide distinct districts with specific use concentrations. Yet, these centers are ultimately designed around people. Unlike existing 400 meters square blocks currently planned and built in China, a fine-grained 100 meters system of small blocks is proposed. Neighbourhoods with small blocks and small streets contribute to a human-scaled, walkable environment. Buildings are built to the street edge, reducing or eliminating setbacks to create intimate and comfortable space shaded by trees and framed by canals.

Integrated Systems

The framework of Shunde is comprised of integrated water, transportation, and wetland park systems; all three expressed as one entity. Waterways define the various areas of Shunde New City, but also become a way of traveling between islands and offer another means of recreation through a water taxi system. A network of water taxi, bus, and monorail stops are located at roughly 500 meters intervals, serving a walkable series of island villages. A monorail loop links village centers; bicycle trails follow the rivers, canals, and boulevards; and high speed rail and ferry offer regional connections to Guangzhou and Hong Kong.

environment - architecture of city
Increasing the Water Edge = A Healthier Ecosystem and City System

Open Space
- Urban Forest
- Wetland

Water Edge conditions
- Natural
- Urban
- Commercial
- Resort

Water-based Approach

Shunde utilizes a water-based approach, maximizing the social and economic value of the site by emphasizing and expanding its inherent qualities of place. The water framework also improves land values, thus attracting creative industries and an educated workforce.

The proposed greenbelt highlights the urban forest and wetlands which make up the interconnecting greenbelt system. Wetlands and bioswales are implemented for water filtration and also help redefine the water edge.

Increased edge conditions and accompanying wetland areas increase opportunity for habitat. The city plan promotes integration of a variety of edge conditions with the open spaces; both natural and urban characters.

环境——城市建筑
Environment—Architecture of city

Challenges
environment
Opportunities

Existing Channalized Rivers
What was once a healthy delta with countless braided river ecologies has been constrained into a three-channelized river system.

Delta and Ecology Restored
A braided system of fine-grained waterways is proposed to increase filtering capacity and reconnect the new planned city of shunde to the river network.

Challenges
economy
Opportunities

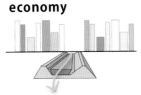

Typical Urbanization / Course Texture
The channeling and consolidation of river systems throughout southern China has destroyed both nature and local culture.

Increasing Edge Condition Creates a Healthy Ecosystem
The plan develops individual islands as pedestrian scaled mixed-use villages that are linked by a proposed environmental infrastructure of green belts, water corridors, rails, trails, and a multilayered transportation system.

Challenges
Culture
Opportunities

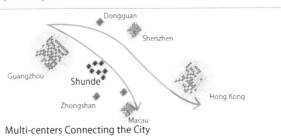

Rapid Urbanization
Unlike existing 400 meter square blocks currently planned and built in China, a fine-grained 100 meter square system of small blocks is proposed.

Multi-centers Connecting the City

挑战——自然与文化的重生
Challenges—Re-born of Nature and culture

the idea - nature and culture re-born

| Ancient Waterways - 1800's | Re-born Delta **environment** | New Multi-centered **economy** | Water Village **culture** |

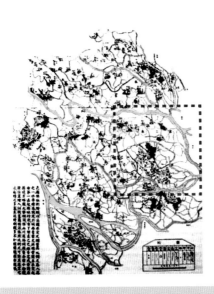   

An unfortunate outgrowth of China's rapid urbanization has been the large scale loss of its natural systems and cultural heritage. These systems served as a buffer against disaster, thus a new model of city is needed where natural and man-made systems become one.

Our idea is to put the delta back, restoring 72-square kilometers of constructed wetlands as the armature for a multimodal city and restore bird and wildlife habitat for the larger Pearl River Delta, while simultaneously expanding flood storage capacity and a lost water-based culture.

The Shunde expansion is comprised of multiple urban centers, each a self-contained unit of residential, retail, office, educational, or civic uses. A civic and cultural center, financial center, office campus, academic campus, and resort provide distinct districts with specific use concentrations.

The historic preservation of the existing village and restoration of water village life adds a valuable link to Shunde's past.

理念——自然与文化的重生
Idea — Re-born of nature and culture

culture - restored water culture

Water Typologies

Traditional Water Village

New Braided Water + Islands

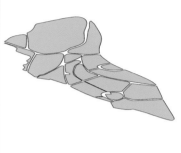

Existing Fish Ponds
Traditional water villages and existing fish ponds form a unique pattern across the landscape.

Adapted Pond Typology

Restore Water Culture
Water village and fish pond typology inspired the new resort water pattern, incorporating clustered residential villas and a village retail center.

environment - urban waterfront park
Urban Run-off Filtered Through Urban Parks

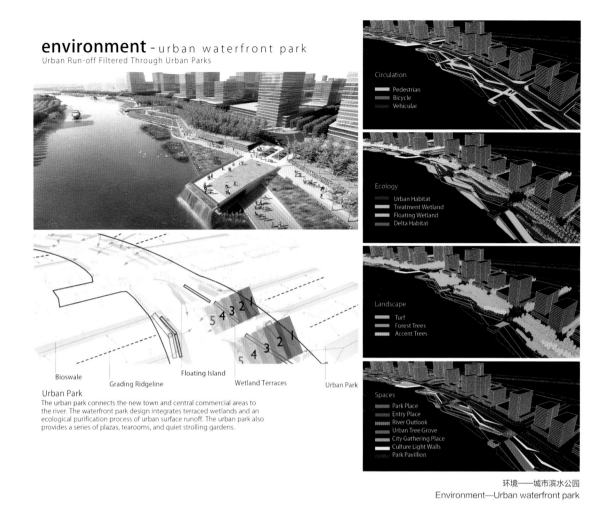

Bioswale Grading Ridgeline Floating Island Wetland Terraces Urban Park

Urban Park
The urban park connects the new town and central commercial areas to the river. The waterfront park design integrates terraced wetlands and an ecological purification process of urban surface runoff. The urban park also provides a series of plazas, tearooms, and quiet strolling gardens.

环境——城市滨水公园
Environment—Urban waterfront park

environment - natural waterfront park
Integrate People Activites Into Nature's Activities

1 Visitors Center + Research institute
2 Aquatic Plant Education
3 Childrens Garden
4 Plant Terraces
5 Amphitheatre
6 Metasequoia Islands
7 Bird Habitat Post
8 Bird Watchtowers
9 River Boardwalk
10 Protected Bird Habitat

Wildlife Park
The wildlife park provides a necessary refuge for Shunde's endangered migratory bird population. The new waterfront design sets aside protected wildlife island preserves as a safe haven and study areas.

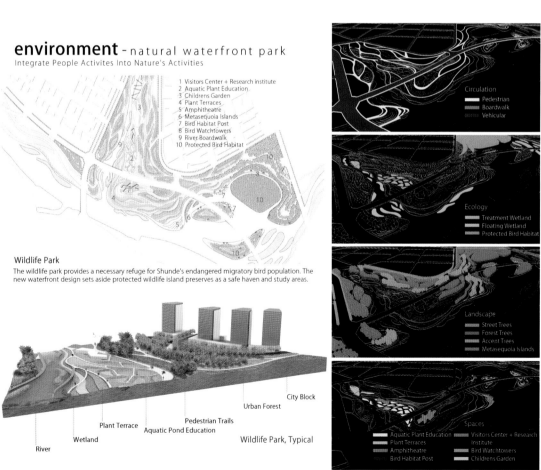

River Wetland Plant Terrace Aquatic Pond Education Pedestrian Trails Urban Forest City Block

Wildlife Park, Typical

环境——自然滨水公园
Environment—Natural waterfront park

078

深圳大运中心水廊道

WATER CORRIDOR OF SHENZHEN UNIVERSIADE CENTER

项目位置：中国广东省深圳市
项目规模：42 ha
设计公司：深圳市北林苑景观及建筑规划设计院有限公司

Location: Shenzhen, Guangdong Province, China
Size: 42 ha
Design Firm: Shenzhen BLY Landscape & Architecture Planning & Design Institute Ltd.

该项目湿塘及雨水湿地合建,设置了7个景观湖,采用堰跌水的连接方式,总容积15万m³,调蓄空间0.7 m。为进一步优化雨水调蓄系统及水质,将铜鼓岭山体引入大运中心,并将临近地块大运自然公园的雨水适量引入大运会体育中心景观湖。同时,优先使用屋面雨水、绿化地雨水,最后采用地面雨水的方向。作为大运会体育中心重要的雨水调蓄措施,为保证水质,屋面和平台雨水经初期弃流后再进行收集,同时尽量将优质雨水(如屋面雨水)收集至景观湖,将略差一点的雨水(如平台雨水)收集至雨水湿地。

This project builds wet ponds and rainfall wetland together with 7 scenic lakes connected by water falls which has a total volume of 150,000 m³ with a regulated capacity of 0.7 m. In order to optimize rainwater regulating storage system and water quality, Tongguling Mountain is introduced into Universiade Center and rainwater of the surrounding Universiade Natural Park is brought into the scenic lake in Universiade Center. In the meanwhile, the sequence is set that roof and Greenland rainwater is used in preference to runoff as the main rainwater regulating measures for the Universiade Center. Roof and platform rainwater is collected after initial rainwater discarding to ensure the water quality and rainwater of high quality (such as roof rainwater) is compiled to the scenic lake as much as possible while rainwater of worse quality (such as platform rainwater) to rainfall wetland.

该项目规划以2011年举办于中国深圳第26届世界大学生运动会为契机,整合了大运场馆中心、运动员村和大运自然公园三块用地,旨在为大运会的举办地营造优美舒适的自然环境,更重要的是关注基础设施建设给自然地的动植物以及当地居民所带来的各种影响,从而高度提升区域生态效益与社会经济价值。

水廊道将大运自然公园、大运中心、大运村三者串联在一起,形成一条绿色通廊。由此形成的绿色网络从大运自然公园向四周延伸,将自然山体、水库、高尔夫球场等纳入其中,形成城市的"绿肺";与之相连的大运中心和大运村作为大运会重要功能区的同时,也将自然环境引入城市。

深圳市属于典型的亚热带海洋性气候,年降水量十分丰富。规划区域内地形多为山地和谷地。降雨后很容易产生大量的山体汇水,并在低洼地形成积水或者湖泊。水廊道通过水体的网络将现状水体和积水区域连接起来,形成水的通道,以此有效地收集降雨,并形成净化水质的湿地,为区域内的生物提供更好的生存环境。同时,项目规划还营造了很多人工的景观水体。它们主要分布在大运中心和大运村中,并为将来的大运会营造丰富的水景效果。自然水通道将人工水体串联在一起,为其提供水源。水廊道顺应地势由西往东延伸开来,形成了自然绿地和城市环境之间的过渡。它不仅是生物廊道的重要支柱,也是城市景观的重要组成部分。

As the 26th World Universiade was held in Shenzhen,china in 2011, this project connects Universiade Center, Universiade Villageand Universiade Natural Park three sites together, hoping to create an elegant and comfortable natural environment for this event and most importantly to highly enhance the ecological and social economic values of the areawith the focus on the impacts of infrastructure construction on the animals, plants and people around.

ConnectingUniversiade Center, Universiade Villageand Universiade Natural Park three sites together, this water corridor forms a greennetwork starting from Universiade Natural Park and extending around to include the natural mountains, reservoir and golf course as the "green lung" of the city. The surrounding Universiade Center and Universiade Village also integrate natural environment into the city while functioning as the main venue for the Universiade event.

Shenzhen is a city of subtropical oceanic climate which has abundant annual precipitation. Considering the planned area mainly consists of mountain land and valley where there would be a lot of mountain catchaent after the rain and ponds in lowland. Water corridor links native waters and ponding area together through water network so as to effectively collect precipitation and form wetlands for water purification, providing a better habitat for wildlife inside the area. In addition, artificial waterscapesare considered in the project planning. Artificial waterscapes are mainly located

鸟瞰图
Aerial view

in Universiade Center and Universiade Village, making a vivid water effect for the coming Universiade event. Natural water channels are connected to and act as water sources for these artificial waterscapes. Water corridor extends from west to east adapted to the topography and become a transition area between natural Greenland and urban environment. It is not only an important pillar for the wildlife corridor, but also a significant component of urban landscape.

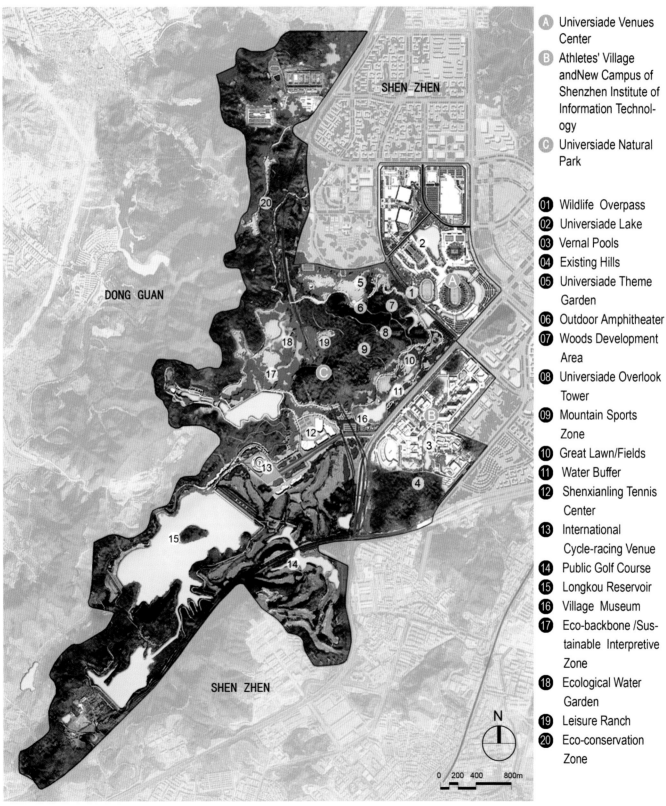

A Universiade Venues Center
B Athletes' Village andNew Campus of Shenzhen Institute of Information Technology
C Universiade Natural Park

01 Wildlife Overpass
02 Universiade Lake
03 Vernal Pools
04 Existing Hills
05 Universiade Theme Garden
06 Outdoor Amphitheater
07 Woods Development Area
08 Universiade Overlook Tower
09 Mountain Sports Zone
10 Great Lawn/Fields
11 Water Buffer
12 Shenxianling Tennis Center
13 International Cycle-racing Venue
14 Public Golf Course
15 Longkou Reservoir
16 Village Museum
17 Eco-backbone /Sustainable Interpretive Zone
18 Ecological Water Garden
19 Leisure Ranch
20 Eco-conservation Zone

总平面图
Site plan

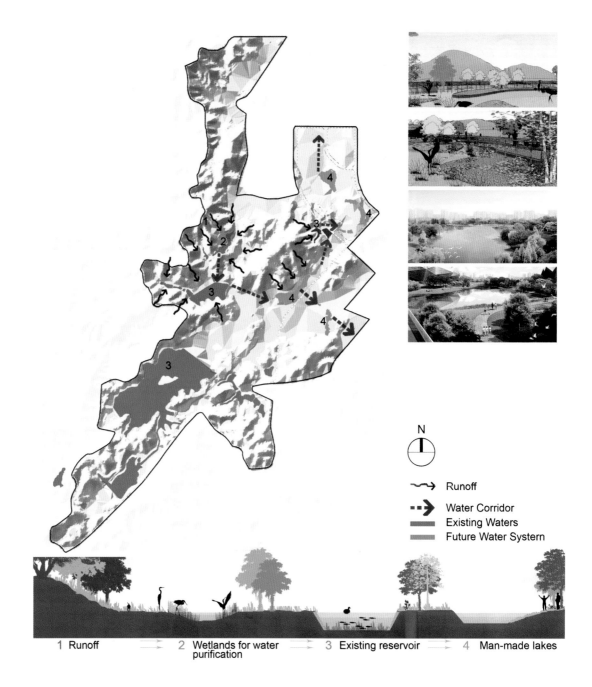

- Runoff
- Water Corridor
- Existing Waters
- Future Water System

1 Runoff → 2 Wetlands for water purification → 3 Existing reservoir → 4 Man-made lakes

水通道
Water channel

透水混凝土
素土壤
砾石层
滤水管
地下车库顶板

雨水

透水铺装
Permeable paving

湿塘（景观湖）
Wet pond(scenic lake)

深圳大运会体育中心鸟瞰图
Aerial view of Shenzhen University Games Sports Center

雨水湿地
Stormwater wetland

绿色屋顶
Green roof

景观湖和多级水坝
Landscape lake and multi-layered dam

透水铺装实景
Permeable paving scenery

086

漂浮的连接：哈尔滨文化中心湿地公园

FLOATING CONNECTION: HARBIN CULTURAL CENTER WETLAND PARK

项目位置：中国黑龙江省哈尔滨市
项目规模：118 ha
设计公司：土人景观

Location: Harbin, Heilong jiang Provice, China
Size: 118 ha
Design Firm: Turenscape

为了管理雨洪径流及净化自来水厂的尾水,哈尔滨市江北区修建了一片城市湿地,并引入亲水休闲设施,包括大部分修建在水面之上的人行栈道和休憩场所。同时,区域内引入部分牲畜维绿地,不仅能开发畜牧产品,还能降低湿地的维护成本,原本"荒野"的自然景观也因此变成一个难得的城市公园,自然而不乏可进入性,一年四季吸引大量市民前往使用,同时为建设水弹性城市作出重要贡献。

挑战与目标

哈尔滨是中国东北地区的重要城市,地处松花江下游,洪泛时有发生。项目场地位于江北新区,当地刚刚修建了一条500年一遇的防洪堤,将原属于江滩的20 ha湿地与主河道切离。由于水源被截断,湿地生境恶化。与此同时,湿地以北的城市建设迅速发展,造成严重的雨洪问题,被污染的雨水排入河道,导致松花江的水质下降。除此之外,新建的自来水厂每天向松花江排放1500 m^3的废水。凡此种种问题,很令当地政府头痛。同时,随着人口规模急剧扩大和城市化进一步发展,人们对公共绿地的渴求也越来越强烈。

设计师提出将这片退化中的湿地改造为雨洪及自来水尾水的净化区,以改善原生的湿地生境,同时作为城市公园,满足日益增长的市民户外游憩需求。研究发现,场地在旱季和雨季的水位高差变化竟达2 m,因此寻求将公共空间与弹性湿地景观相结合便成为项目成功与否的关键。此外,如何管理如此大面积的城市公共空间也是一个难题。因为用不了多久,修复后的水生和湿生植被将滋生繁衍,市民也将很难在一年四季内都能进入这片湿地。

项目的目标是构建水弹性湿地公园,使之成为生态基础设施的有机组成部分,并用于净化雨洪和自来水厂排放的废弃尾水。在这一过程中,湿地重现生机。此外,景观设计师认为,有限的设计干预措施是实现项目目标的最佳手段,在恢复自然和发挥其雨洪管理即生态水净化作用的同时,打造一片低维护的城市绿地。

设计方案:功能性湿地、弹性公园和景观牧场

净化雨洪的湿地:湿地公园四周将修建一系列生态洼地,南部文化中心及北部城区产生的雨洪径流将汇集于此并得到净化。这个湿地系统在雨季日均可处理2万m^3的水量,在雨洪流入湿地中心地带之前,生态洼地降低水流中的泥沙含量,拦截营养物及重金属。除此之外,附近自来水厂排放的1500 m^3的尾水也将首先进入这片湿地进行净化,以免污染河道。

漂浮的连接与水弹性公园:通过配置适应性植被,使水岸绿化可应对旱季和雨季高达2 m的水位差。水位线以上的植被采用放任自然的管理,较高的地带则在配置的乡土树丛之间种植人工草甸。在生态洼地的修建过程中,尽可能减少工程量,将开挖和回填过程结合在一起,同时尽量避免坡地整理的土方工程,以最大限度地保留原有的树木和地被。每年的水位波动使水域边缘出现泥泞、脏乱、难以靠近的地带,若将湿地建设成市民终年可以前往活动的公共空间,显然需要处理这个难题。设计师的办法是搭建离岸木栈道,使人行空间与地面脱离了湿地边缘。6 km的栈道系统连接13个休憩平台。此外,利用当地透水的火山沙,在高地上铺设出生态友好型的人行道,与湿地上的栈道共同形成一个连续的步行网络,穿梭于树丛和草地之间,极大地丰富游客的景观体验。

畜牧景观与低维护需求:大多数时候,公园受自然过程控制,而非人工管理,因此湿地和地面的植被很容易疯长,从而形成杂乱、难以进入的自然景观。设计师提出将牲畜引入较高的地带进行放牧,以抑制植物疯长,湿地景观可得到低成本的维护,公园也可全天候对游客开放,同时具有生产功能,并为游客提供丰富的景观体验。

结论

通过以上这些景观设计策略,原先持续恶化的湿地生境和无人问津的城市边缘地带,已成功地被改造成一个具有功能的湿地公园,可用于调节雨涝,净化城市地表径流以及自来水厂排放的尾水。在弹性公园里,采取最小限度的干预措施修建了木板道和休憩场所,轻轻地触摸大地,亲近自然而不破坏自然,满足市民休闲游憩需求的同时让自然休养生息。

游憩系统是完全无障碍的,使各种身体条件的人都能够接触自然。
The integrated pedestrian network of walking and resting facilities are detached from natural habitats and are universally accessible, allowing people of all abilities to have intimate touch with the marshy nature.

木板道和桥梁与地面和湿地的边缘相分离，因此对场地内的自然环境只有最低程度的影响，否则它们可能显得杂乱。这些文化性构筑物创造了一种不干扰自然过程和形态的新模式。

Floating above the ground and the wetland edge, the boardwalks and bridges are minimum interventions that enframe nature, which might otherwise be viewed as messy. These cultural artifacts create a new form without disrupting the natural processes and patterns.

在树林和湿地边缘之间打造了由众多乡土物种构成的片片草地，生命力旺盛的花卉在此绽放。每年的花季，当人们在密集树丛围绕的高地与花丛之间相遇时，彼此惊喜万分。

Native meadows were designed between the forest groves and the wetland edge. These are composed of self-seeding flowers creating annual surprises when they are encountered within the enclosing densely forested uplands.

The urban storm-water is drained into the river

The tail water from the city water supply plant drained into the river causing pollution

The wetland habitat is deteriorating due to the new flood wall that cut off the wetland

旧场地的照片反映出这块场地的恶劣环境。一道修成不久的防洪墙把场地与主河道分隔开。来自一座自来水工厂的城市雨水径流和尾水导致了河流污染。受到损害的生态系统和环境污染对湿地公园的设计形成了挑战，同时也带来了灵感。

The aerial photographs along with photos of the pre-exiting site reveal the deteriorating condition of the site due to its isolation from the main river resulting from the recent flood wall. Urban storm water runoff and tail water from a water supply plant caused pollution to harm the river. The degraded ecosystem and pollution challenged and inspired the design of the wetland park.

An urban wetland was created as a solution to manage storm-water runoff and tail water from water supply plant. A network of water adaptive recreational uses were introduced, including pedestrian paths and resting places, which are largely detached from the ground. Cattle were also introduced to create an agricultural husbandry that is productive while creating a low maintenance solution to make the otherwise "messy" nature into a cherished urban park that is intensively used by urban residents year round.

Challenges and Objectives

Harbin is one of the major Chinese cities in northeast area of the country. The city is subject to flooding due to its location at the lower reach of the Songhuajiang River. An existing 500-year flood control wall along the riverbank cuts off a 200-hectare wetland from the main river. The wetland habitat was deteriorating due to the lack of water supply. Meanwhile, the rapid urban development north of the site caused severe storm-water inundation while the discharge of the contaminated storm water caused the water quality of the river to decline. In addition, tail water from a newly built water supply plant dumps 1,500 cubic meters of contaminated wastewater into the Songhuajiang River. The sum of these development and pollution inputs created worrisome problems for the local community. At the same time, public space is fervently desired due to the dramatic increase of urban population and development.

The initial understanding of the site was to turn the isolated wetland into a major park as well as a storm water and wastewater remediation area that would enhance native wetland habitat. But thorny issues emerged as the result of further study. The landscape architect discovered that the seasonal change of the water table is as much as 2 meters between dry and wet seasons, which thwarts the intention to combine public spaces with a resilient wetland landscape. In addition, such a large public space would be difficult to manage since the restored native vegetation would soon become too messy and wild to allow access and used by the people year round.

The design objective was to fashion a water resilient wetland park, which functions as an integral ecological infrastructure that remediates storm water and waste tail water from the water plant. In the process, the waste water could rejuvenate the wetland habitat. Furthermore, the landscape architect determined that limited design interventions would best serve the project objectives and transform the wetland into an accessible public space. A final challenge was establishing a low maintenance program for the park.

Design Strategies—Functional Wetland, Resilient Park and Landscape Pasture

A working wetland for storm water remediation: A series of bio-swales were designed for the periphery of the wetland park to catch and filter the storm water runoff from the new development of the cultural center at the south and city at the north. Each day during the raining season, up to 20,000 cubic meters (706,000 ft3) of storm water on average will drain into the wetland filtration system. The bio-swales will reduce the sediment, suspended solids and heavy metals in the storm water before it flows into the central part of the wetland. In addition, 1,500 cubic meters of tail water from the nearby water supply plant is retained and treated in the wetland to avoid polluting the river.

Floating connection and water resilient park: Adaptive vegetation was designed to accommodate the two-meter annual fluctuation of the water table. In the low land, vegetation was left to respond to the process of natural succession. On the higher land, stretches of semi-natural meadow were sewn between the planted native tree groves (Betula platyphylla and Fraxinus mandschurica). Except for the minimum cut-and-fill earthwork necessary to create bio-swales, little grading was planned so that all of the existing trees and ground cover could be preserved. The annual fluctuation of the water table creates an inaccessible muddy and messy water edge, and poses a huge challenge to making the wetland into a public space that would be accessible year round. The design solution was a network of boardwalk (and bridges) detached from the ground and wetland edge. Altogether 6 kilometers of boardwalk and bridges were built, linking 13 platforms and pavilions. In addition, an eco-friendly network of pedestrian paths (using permeable volcanic sand from this region) is built on the raised ribbons of land that penetrate the groves and meadows, providing a rich experience for visitors.

Husbandry landscape and low maintenance: Since the park is largely under the control of natural, rather than human, processes, it is likely to become overgrown with seasonal wetland vegetation and ground cover, forming a wild and messy naturalized landscape. Cattle were introduced to maintain low vegetation on the high land zone so that the landscape will be accessible and usable as a park. A second benefit is the production of food and an enriched landscape experience for the city dwellers that have long been separated from agriculture.

Conclusion

By these landscape strategies, the former deteriorated wetland habitat and a neglected peri-urban site has been successfully transformed into a working wetland that remediates urban storm water runoff and waste tail water from the water supply plant. An adaptive water resilient park was created through the minimum intervention of a network of boardwalk and resting places, which enable people to be close to nature but without disturbing nature. Thus, the recreational use of a park for people is fulfilled, and the nature can recuperate.

13个平台均抬离地面,以便将对乡土生物的影响降至最低。为了与周边景观相呼应,每一个平台都经过精心设计,在繁茂的植被中形成令人惊喜的对比。这便是运用最小化干预手段让杂乱无章的大自然变得更可观的典型范例。

Each of the 13 resting platforms are detached of the ground to impose minimum impact on the native habitat. Each of them is uniquely designed in response to the local setting, and creates a contrasting surprise in the overgrown landscape. It is a minimum strategy to order and enframe the messy nature.

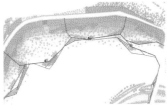

The water adaptive board walk and platforms detached from the ground and wetland edges

The site plan

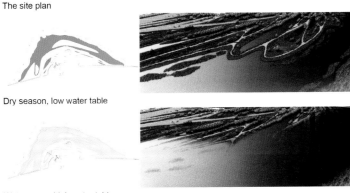

Dry season, low water table

Wet season, high water table

Bio-swale that catches storm-water runoffs

场地地图及设计理念：这是一个拥有适应性木板小路和人行天桥的水弹性公园。道路和桥梁与随季节而发生变化的地面和湿地边缘相隔离。园中设计了生态沼泽，以缓和雨水径流。
Site map and design concept: A water resilient park with adaptive elevated boardwalkspedestrian bridges and platforms, which are detached from the seasonally changing wetland edges. The bio-swales are used to remediate stormwater runoff.

设计师战略性地将13个平台和亭子沿6 km长的木板路和天桥进行安置。这些休闲空间提供了观赏湿地公园和城市景观的绝佳角度。
The 13 platforms and pavilions are strategically located along the 6 kilometers of boardwalks and bridges. These resting places provide great views of the wetland park and city landscape.

在建成湿地公园之后，这张北部全景照片包含了背景中新的城市开发。木板路和桥梁体系与地面和水岸分开，漂浮在恢复了原始形态的湿地之上
A northern panorama of the wetland park with new urban development in the background: The network of boardwalks and bridges detached from the ground and water edges, floats above and runs through native communities, allowing intimate connection with nature without interrupting its processes

漂浮的连接：茂盛的植物吸引了各类野生动物，漂浮于湿地之上的步行道和桥梁网络，使游客在不打扰动物活动的情况下与自然亲密接触（图片中游客在观赏鸟和钓鱼）
Floating Connection: The flourishing regenerated wetland attracts diverse wildlife while the floating board walks and platforms allowing visitors to have an intimate contact with nature without interrupting its processes. (visitors watching birds and fishing at the background of the image).

在人行天桥上行走是一种愉快的体验,行人能够与自然产生亲密的联系。夏日盛开的荷花吸引着大量游人,步行桥的设计使他们可以近距离地观赏水景。
Walking on the pedestrian bridges provides a pleasant experience that allows intimate contact with the nature. The blossoming wild lotus flowers in the summer attract thousands of visitors, which are otherwise invisible and untouchable due to their remote siting in the deep marsh.

高架木板路和桥与水岸相分离,供市民游赏。它们组合成片片景观,并在不同的季节提供欣赏美丽而脆弱的水景观的路径。
Urban residents enjoy platforms built into the elevated boardwalks and bridges above the water edge. They organize the landscape and provide access to the beauty of the fragile aquatic landscape in every season.

步行道使用当地出产的渗透性火山砂,形成一个生态友好型网络。道路建造在高地上,穿行于树林和草地之中。它们为参观者提供了丰富多彩的景观体验。
An eco-friendly network of pedestrian paths (using the permeable volcanic lava sand from this region) is built on the high ground that penetrates the groves and meadows. They create a rich experience for visitors.

一系列生态洼地和池塘作为吸收开发区雨水的缓冲带。
A series of bio-swales and ponds are designed as buffers to remediate the storm water runoff from the development area.

简约式的设计元素如树林、草地和当地火山砂铺就的步道,有效地将公园转变为景色宜人且可提供丰富体验的场所,而非通常所见开放式公园。
Minimalist design elements such as including tree groves, meadows and colored accesses (pave with local lava sand) have effectively transformed the otherwise open and generic landscape into pleasant spaces and places accommodating various recreational activities.

在新建的公园中生长着繁茂簇新的植物,为孩子们的嬉戏玩耍提供了美丽的空间。
Children have the opportunity to play and interact with the lush vegetation via the newly created park.

096

弹性景观：金华燕尾洲公园

RESILIENT LANDSCAPE: JINHUA YANWEIZHOU PARK

项目地点：中国浙江省金华市
项目规模：26 ha
设计公司：土人景观

Location: Jinhua City, Zhejiang Province, China
Size: 26 ha
Design Firm: Turenscape

建成景观鸟瞰图（旱季）。得益于雨季洪水带来的泥沙沉积，防洪梯田更加有效。禾本科植被茂盛地生长。游人在其中欢快游憩。（西望图景，2014年11月）

This is the aerial view of the park during the dry season. The lush tall grasses cover the terraces on the embankment. The terraces are enriched by silt deposit during the flood season. Visitors enjoy a good time here. (towards the west, photo in November 2014)

该项目规划通过一个实验性工程,探索了如何通过设计实现景观、社会和文化的弹性发展,重点探索了如何与洪水为友,营造适应性防洪堤、适应性植被和百分之百的透水铺装,建设适应于多方向人流的步行和桥梁系统及社区纽带,实现景观的生态弹性。灵动的流线设计语言,将场地上原有的流线型建筑、季节性的水流和川流不息的人流有机地编织在一起,溶解场地,解决了瞬时人流和日常休闲空间的使用矛盾,营造了富有弹性的体验空间和社会交往空间,实现了景观的社会弹性;项目规划从当地富有历史和文化意味的"板凳龙"传统舞龙习俗中获得灵感,设计了一条富有动感、与洪水相适应的步行桥,将被河流分割的两岸城市连接在一起,并使河滩变成富有弹性的可使用景观,形成了被称为"金华市最富诗意的景观",将断裂的文脉串联在一起,强化了地域文化的认同感和归属感,实现了景观的文化弹性。

挑战与目标

在隔江相望的城市包围下,燕尾洲已经成为金华这一具有100万人口的繁华都市中唯一尚有自然兼葭和枫杨的芳洲。义乌江和武义江在此交汇而成婺江(金华江)。洲的大部分土地已经被开发为金华市的文化中心,现建有中国婺剧院,为曲线异形建筑,洲的两侧堤岸分别是密集的城市居民区和滨江公园,但由于开阔的江面阻隔,市民难以到达和使用洲上的文化设施。留下的洲头共26 ha的河漫滩,其中部分因采砂留下坑凹和石堆,地形破碎,另一部分尚存茂密植被和湿地,受季风性气候影响,每年受水淹没,形成了以杨树和枫杨为优势物种的群落,是金华市中心唯一留存的河漫滩生境,为多种鸟类和生物提供庇护,包括当地具有标志意义的白鹭。

因此,设计面临四大挑战:①如何在供市民使用的同时,保护城市中心仅有的河漫滩生境,给人口稠密的城市留下一片彼岸方舟?②如何应对洪水,是高堤防洪建一处永无水患的公园,还是与洪水为友,建设与洪水相适应的水弹性景观?③如何处理与现有的异形建筑体和场地的关系,形成和谐统一的景观整体?④如何连接城市南北,给市民提供方便使用的公共空间,并强化城市的社会文化认同感?

弹性设计策略

1. 保护自然与修复生态的适应性设计

由于长期采砂,场地坑洼不平,地形破碎。针对这一特点,设计通过最少的工程手段,保留原有植被;在原有坑塘和高地基础上,稍加整理,形成滩、塘、沼、岛、林等生境,以培育丰富的植被景观。在此基础上,结合各类生境的特点进行植被群落设计,重点补充能优化水质的水生藻类、沉水、浮水植物、能为鸟类和其他动物提供食物的浆果类植物以及具有季相变化的乡土树种等。由此,完善和丰富了场地中的植被和生物多样性。

2. 与水为友的弹性设计

金华地处中国东部亚热带地区,受强烈的海洋季风性气候的影响,旱、雨季分明,雨季常受洪水之扰。同时,为了争取更多的便宜土地进行城市建设,大量河漫滩被围合开发。两江沿岸筑起了水泥高堤以御洪水,隔断了人与江、城与江、植物与江水的联系。同时,江面缩窄,也使洪水的破坏力更加强大。为保护沙洲不被淹没,当地水利部门已经在燕尾洲的部分地段,分别修建了20年一遇和50年一遇的两道防洪堤,破坏了沙洲公园的亲水性。该设计不但将尚没有被防洪高堤围合的洲头设计为可淹没区,同时,将公园范围内的防洪硬岸砸掉,应用填挖方就地平衡原理,将河岸改造为多级可淹没的梯田种植带,增加了河道的行洪断面,缓减了水流的速度,缓解了对岸城市一侧的防洪压力,提高了公园邻水界面的亲水性。梯田上广植适应于季节性洪涝的乡土植被,梯田挡墙为可进入的步行道网络,使滨江水岸成为生机勃勃、兼具休息和防洪功能的美丽景观。每年的洪水为梯田上多

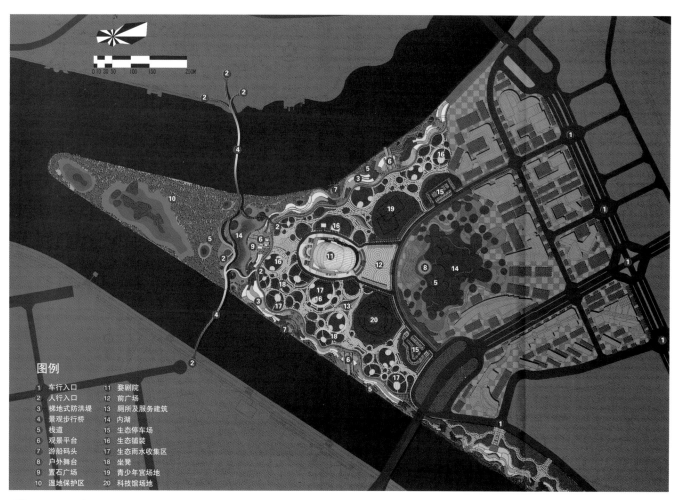

弹性景观总平面图:金华燕尾洲公园
The site plan for resilient landscape: Yanweizhou Park, Jinhua.

图例
1 车行入口　11 婺剧院
2 人行入口　12 前广场
3 梯地式防洪堤　13 厕所及服务建筑
4 景观步行桥　14 内湖
5 栈道　15 生态停车场
6 观景平台　16 生态铺装
7 游船码头　17 生态雨水收集区
8 户外舞台　18 坐凳
9 置石广场　19 青少年宫场地
10 湿地保护区　20 科技馆场地

年生蒿草带来充足的沙土、水分和养分,使其茂盛地繁衍和生长,且不需要任何施肥和灌溉。梯田河岸同时将来自陆地的面源污染和雨洪进行滞蓄和过滤,避免对河道造成污染。该项目尽管只有一段微不足道的生态防洪区域,但可作为婺江流域河道防洪设计的样板,供借鉴和推广。

除了水弹性的河岸设计外,场地内部也采用百分之百的可下渗覆盖,包括大面积的沙粒铺装作为人流的活动场所,与种植结合的泡状雨水收集池,以及用于车辆交通的透水混凝土道路铺装和生态停车场,实现了全场地范围内的水弹性设计。

3. 连接城市与自然、历史与未来的弹性步桥

横跨三江六岸的富有弹性和动感的步行桥,连接城市的南北两大组团,以及城市与江洲公园。步行桥的设计以金华当地民俗文化中的"板凳龙"作为灵感来源。这是金华当地特有的春节龙舞,每家每户搬出自己的板凳,连接在一起形成一条长龙,敲锣打鼓蜿蜒在田埂上,全村老少喜气洋洋地跟在其后。"板凳龙"不仅仅是一种狂欢的舞蹈,更是社区和家族的纽带,它灵活机动,具有社会文化认同的象征意义。彩桥因地势盘旋扭转,富有弹性,结合缓坡设计巧妙地解决竖向高差,其中连接城南-城北的主要桥体在200年一遇的洪水范围之上,以保证在特大洪水时都能同行,而其中与燕尾洲公园连接的部分,则可以在20年一遇的洪水中淹没,以适应洪水对沙洲湿地的短时淹没。步行桥飘忽燕尾洲头的植被之上,使游客可在城市之中近距离地触摸真实的自然。色彩上,采用具有民俗特征和喜庆炽烈的红黄交替,同时结合晚间灯光和照明功能,流畅绚丽、便捷轻盈。桥梁总长超过700 m,其中跨越义乌江、武义江段分别为210 m和180 m。步行桥全线采用钢箱结构,桥梁主线宽5 m,匝道宽4 m,桥面采用环保材料竹木铺设,发光栏杆则选用新型的透光玻璃钢材料。

这座桥的建成大大缩短了城南和城北的步行交通距离,并将两岸绿廊和多个公园串联在一起。步行桥已被正式命名为"八咏桥",以纪念历史上咏叹金华四周景观的八首诗歌。无论从其对水的适应弹性,还是对来自各个方向的人流疏导及使用强度的适应性,抑或其作为连接城市与自然、历史与未来的黏结性,"八咏桥"都可称为一座富有弹性的桥。倘徉在飘舞的"八咏桥"上,看金华城市及四周的连绵山峦、蜿蜒而来的河流与川流不息的人,诗意油然而生。难怪当地市民称其为"最富诗意的桥"。

4. 动感流线编织的弹性体验空间

圆弧形的大型建筑(金华婺剧院)给场地空间和形态设计提出了挑战,即如何在营造弹性空间的同时满足瞬时集散和平时游人的空间需求和体验,如何营造宜人的环境,将游憩空间、防洪及巨型建筑与江岸都纳入其中等。该设计在形式语言上大胆运用了流线,包括河岸梯田和流线型种植带、流线型地面铺装、流线型道路、空中步道和跨河步行桥。在流线的铺装纹理基底上,设置圆弧形的种植池,里面种满水杉或竹丛,色彩鲜艳的圆弧形座椅作为边界。圆形种植区是场地雨水的收集区,如雨滴落在水面上泛起的圆形水波。流线与圆弧形线条和形体既是建筑与环境相融合的语言表达,更是水流、人流和物体势能的动感体现,形式与内容形成了统一,环境与物体得以和谐共融,营造了极富动感的体验空间。

结论

经过两年的设计和施工,燕尾洲公园的建设取得了巨大的成功。2014年5月开园后,万人空巷,游人如织。目前,燕尾洲公园已经成为金华市的一张新名片。

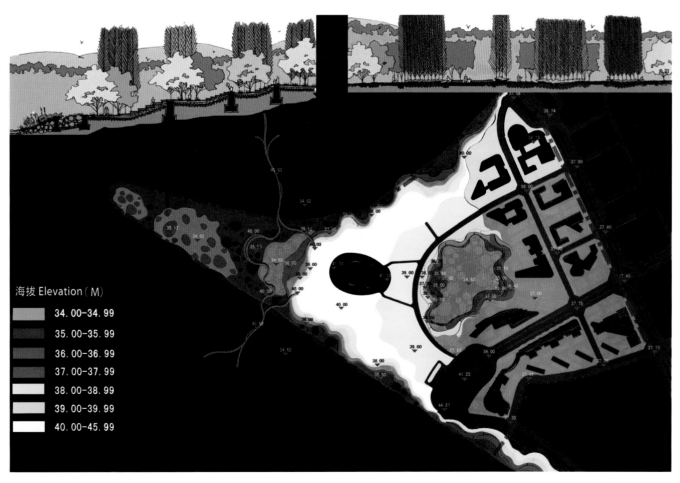

适应于洪水的场地标高设计。两侧的梯田取代原有的水泥防洪堤,场地内部为百分之百的可透水覆盖,包括人行区的砂石、与种植相结合的生态集雨区、车行区的透水水泥铺装和生态停车场
Grading Plan and Section: flood walls are removed and a cut-and-fill strategy is used to create terraces to make the site cooperate with flood. The design is 100% permeable. The surface includes gravel surface for pedestrians, bio-swales for planting, and permeable concrete for automobile use.

Water resilient terrain and plantings are designed to adapt to the monsoon floods; a resilient bridge and paths system are designed to adapt to the dynamic water currents and people flows. The bridge and paths connect the city with nature and connect the past to the future; resilient spaces are created to fulfill the need for temporary, intensive use by the audience from the opera house, yet are adaptable for daily use by people seeking intimate and shaded spaces. The river currents, the flow of people, and the gravity of objects are all woven together to form a dynamic concord. This is achieved through the meandering vegetated terraces, curvilinear paths, a serpentine bridge, circular bio-swales and planting beds, and curved benches. The project has given the city a new identity and is now acclaimed as its most poetic landscape.

Challenges and Objectives

In the urban heart of Jinhua, a city with a population of over one million, one last piece of natural riparian wetland of more than 26 hectares (64 acres) remains undeveloped. Located where the Wuyi River and Yiwu River converge to form Jinhua River, this wetland is called Yanweizhou, meaning "the sparrow tail". Beyond this tail, riparian wetlands have already been eliminated by the construction of an organically shaped opera house.

Before the Yanweizhou Park project was implemented, the three rivers, each of which is over 100 meters wide, divide the densely populated communities in the region. As a result of this inaccessibility, the cultural facilities, including the opera house and the green spaces adjacent to the Yanweizhou, were underutilized. Most of the riparian wetland has been fragmented or destroyed by sand quarries, and is now covered with secondary growth, which was dominated by poplar trees (Populus Canadensis) and Chinese Wingnut (Pterocarya stenoptera) that provide habitat for native birds like egrets.

The site conditions posed four major challenges to the landscape architect: 1) How can the remaining riparian habitat be preserved while providing amenities to the residents of the dense urban center? 2) What approach to flood control should be used, prevention with a high, concrete retaining wall or cooperation by allowing the park to flood? 3) How can the existing organically shaped building be integrated into the surrounding environment to create a cohesive landscape that provides a unique experience for visitors? 4) Finally and most importantly, how can the separated city districts be connected to the natural riparian landscape to strengthen the community and cultural identity of Jinhua?

Design Strategy: Resilient Landscape

1. Adaptive tactics to preserve and enhance the remnant habitats

The first adaptive strategy was to make full use of the existing riparian sand quarries with minimum intervention. In this way, the existing micro-terrain and natural vegetation are preserved, allowing diverse habitats to evolve through time. The biodiversity of the area was adapted and enhanced through the addition of native wetland species. This enrichaent, particularly of species that provide food for birds and other wildlife, increases biodiversity.

2. Water resilient terrain and planting design

Due to its monsoon climate, Jinhua suffers from annual flooding. For a long time, the strategy to control flooding was to build harder and taller concrete walls to yield cheap land for urban development. These walls along the riverbanks and riparian flood plains severed the intimate relationship between the city, the vegetation, and the water, while ultimately exacerbating the destructive force of the annual floods.

The preexisiting site (2011)　　　　　Before (2011)　　　　After (2014)

场地现状及建成前后对比。两侧水泥防洪堤被改造成梯田式生态护堤，蜿蜒的空中步行桥将被分割的城市连为一体。
This is pre-existing conditions (before) and transformations (after). This existing site was a riparian wetland ruined by sand quarries and concrete flood walls. The resilient design strategy dramatically transformed the site by eco-friendly embankment and making the site accessible and connecting the segregated city.

Dry season 旱季

Flood season 雨季

与洪水相适应的栈道（5年一遇的高度），并与梯田的田埂系统相结合，形成一个空间体验系统，使游客与自然亲密接触。（西南图景）
The flood adaptive boardwalk integrated the path system with the terraces. This path affords visitors an intimate naturalistic experience over the riparian vegetation. The elevated boardwalk is just above the five-year flood level. (toward the southwest)

去掉高高的水泥防洪堤，通过就地平衡土方的"填-挖"策略，建立梯级生态护坡，形成洪水缓冲区。（西北图景）
The beautiful terraced embankment was built by removing the concrete flood wall and through a cut-and-fill strategy that balances the earthwork on-site.

Following this formula, hard high walls have been built, or were planned to be built, to protect the last patch of riparian wetland (Yanweizhou) from the 20-year and 50-year floods. These floodwalls would create dry parkland above the water, but destroy the lush and dynamic wetland ecosystem. Therefore, the landscape architect devised a contrasting solution and convinced the city authority to stop the construction of the concrete floodwall as well as demolish others. Instead, the Yanweizhou Park project "makes friends" with flooding by using a cut-and-fill strategy to balance earthwork and by creating a water-resilient, terraced river embankment that is covered with flood adapted native vegetation. Floodable pedestrian paths and pavilions are integrated with the planting terraces, which will be closed to the public during the short period of flooding. The floods bring fertile silt that is deposited over the terraces and enrich the growing condition for the tall grasses that are native to the riparian habitat. Therefore, no irrigation or fertilization is required at any time of the year. The terraced embankment will also remediate and filtrate the storm water from the pavement above. Although the design and strategies address only a small section compared to the hundreds of kilometers of river embankment, the Yanweizhou Park project showcases a replicable and resilient ecological solution to large-scale flood management.

In addition to the terraced river embankment, the inland area is entirely permeable in order to create a water resilient landscape through the extensive use of gravel that is re-used material from the site. The gravel is used for the pedestrian areas; the circular bio-swales are integrated with tree planters, and permeable concrete pavement is used for vehicular access routes and parking lots. The inner pond on the inland is designed to encourage river water to infiltrate through gravel layers. This mechanically and biologically improves the water quality to make the water swimmable.

3. A Resilient Pedestrian Bridge Connects City and Nature, Future and Past

A pedestrian bridge snakes across the two rivers, linking the parks along the riverbanks in both the southern and northern city districts, and connecting the city with Yanweizhou Park within the river. The bridge design was inspired by the local tradition of dragon dancing during the Spring Festival. For this celebration many families bind their wooden benches together to create a long and colorful dragon that winds through the fields and along narrow dirt paths. Musicians sound gongs and beat drums, to the accompaniment of singing, dancing and yelling by villagers, young and old. The Bench Dragon is flexible in length and form as people join or leave the celebration. The dragon bends and twists according to the force of human flow. Like the bench dragon in the annual celebration, the "Bench Dragon Bridge" symbolizes not only a form of celebration practiced in Jinhua area, but is a bond that strengthens a cultural and social identity that is unique to this area . As water-resilient infrastructure, the new bridge is elevated above the 200-year flood level, while the ramps connecting the riparian wetland park can be submerged during the 20-year and larger floods. The bridge also hovers above the preserved patch of riparian wetland and allows visitors an intimate connection to nature within the city. The many ramps to the bridge create flexible and easy access for residents from various locations of the city in adaptation to the flow of people. The landscape architect designed the bridge to reinforce the festive, vernacular tradition, but also as an art form with a bold and colorful combination of bright red and yellow tones that are strengthened by night lighting. All together 700 m long, the bridge is composed of a steel structure with fiberglass handrails and bamboo paving. The main bridge is five meters wide, with four-meter wide ramps. This bridge is officially named Bayong Bridge (Bridge of Eight Chants), after eight famous poems written in ancient times about landscapes surrounding the site. It is truly a resilient bridge that is adaptive to river currents and the flows of people while binding city and nature, future and past.

4. Resilient Space for Dynamic Experience

The large oval opera house posed significant challenges for the landscape architect. First the building shape tends to repel rather than embrace the

建成景观鸟瞰图（雨季），20年一遇的洪水淹没的实景。即便如此，步行桥仍然维护两岸的有效通行。（西望图景，2014年5月）
This is the aerial view of the park during the monsoon season showing a 20-year flood and testifying to the flood resilient design. The uninterrupted connection of the city is ensured by through the bridge. (towards the west, photo in May 2014)

user and landscape. Therefore, the first challenge was devising innovative forms that would welcome and embrace the visitors. Secondly, the area near the building needed to accommodate the large opera audience as well as offer intimate spaces and ample shade. Finally, the designers were challenged with the problem of how to integrate the singular flood-proof big object into the floodable, riparian waterfront. The design uses curves as the basic language, including the curvilinear bridge, terraces and planting beds, concentric paving bands of black and white, and meandering paths that define circular and oval planting areas and activity spaces. The spatial organization and design forms establish an extensive paved area for a large audience during the events at the opera house. However, the forms and the inclusion of alcoves create places for the individual, couples and small groups. The dynamic ground of the pavement and planting patterns define circular bio-swales and planting beds, densely planted with native trees and bamboo, bound by long benches made of fiberglass. The circular bio-swales and planting patches resemble raindrop ripples on the river. These curves and circles are the unifying pattern language that integrates the building and the environment into a harmonious whole. The reverse curves simultaneously refer to the shape and scale of the building, while forming a contrasting shape that is human in scale and enclosed for more intimate gatherings. They also reflect the weaving of the dynamic fluxes of currents, people and material objects that together create a lively pleasant and functional space.

Conclusion

The project is a proven success. And now, the Yanweizhou Park has created a new identity for the city of Jinhua.

从当地民俗的舞龙——"板凳龙"中获得设计灵感,跨过两江的步行桥(八咏桥)蜿蜒多姿,它不仅是一条连接通道,更是体验的场所,吸引着大量的游客和居民,每天平均有4万余人使用该桥。它强化了市民对乡土文化的认同感和归属感。(晨景,2014年)

Inspired by the vernacular Bench Dragon Dancing, the iconic Bayong Qiao bridge is more than a connecting infrastructure, it attracts thousands of residents and tourists, and over forty thousand people visit the bridges daily. It recovers the vernacular cultural identity of the city. (early morning scene, 2014)

去掉水泥防洪堤后,河水可通过卵石层过滤净化后进入堤内低地,原有的采沙坑变成了可以游泳和戏水的内湖。(戏水,2014年夏天)

The inner pond on the inland is designed to allow water to infiltrate from the river through the gravel layers that make the otherwise dirty river water swimmable. (kids playing in the pond, summer, 2014)

通过多个连接坡道的设计,舞动的步行桥与来自不同方向的人流相适应,长期被河流分割的社区串联在一起,为家庭成员和素不相识的市民开辟了交流的场所,他们可以每天在这里跑步和聊天。

With multiple ramps following the flow of people, the Bayong Qiao bridge is more than a physical link, but a social connection that helps to build the community by creating a gathering space for families and residents who walk, jog and chat together on the bridge.

适应性植被茂盛的生长，平时为游客提供美丽的体验空间（东望）。
The terraced has created a flood resilient zone that allows people to enjoy the lush tall grasses adaptive to the seasonal floods (towards the east).

圆形生态渗水区与色彩亮丽的板凳相结合，将大型文化演艺建筑的户外场地转变为亲切、阴凉的户外体验场所。集水区内种植与水涝相适应的水杉林。
Circular spaces dissolve the extensive paved area needed when the opera house is in use and provides intimate shaded spaces for daily use. Colorful benches encircle the bio-swales planted with native water adaptive species such as Chinese Redwood.

场地内铺设百分之百的可透水铺装，包括步行区的沙粒铺装、集雨区的生态种植池、车行区的可透水水泥铺装和生态停车场。

The surface of the inland area is hundred percent permeable. Generated from on-site materials, gravel is recycled to create pedestrian surfaces. Bio-swales are integrated with tree planters, and permeable concrete pavement is used for automobile routes traffic use and parking lots.

色彩亮丽的板凳
Colorful benches

高挑的平台和凉亭，地处200年一遇的洪水范围之上，让游人近能俯瞰内湖景观，远能眺望城市。
This pavilion provides a dramatic viewpoint as it extends above the 200-year flood level. The pavilion features a detailed view of the pond water feature and expansive views of the river, the city and the Bayong Qiao bridge.

106

南湖：农业村落的重塑再造

NANHU: REIMAGINING THE AGRICULTURAL VILLAGE

项目位置：中国浙江省嘉兴市
项目规模：1100 ha
设计公司：SWA集团
翻　　译：张旭

Location: Jiaxing, Zhejiang Province, China
Size: 1,100 ha
Design Firm: SWA Group
Translated by Zhang Xu

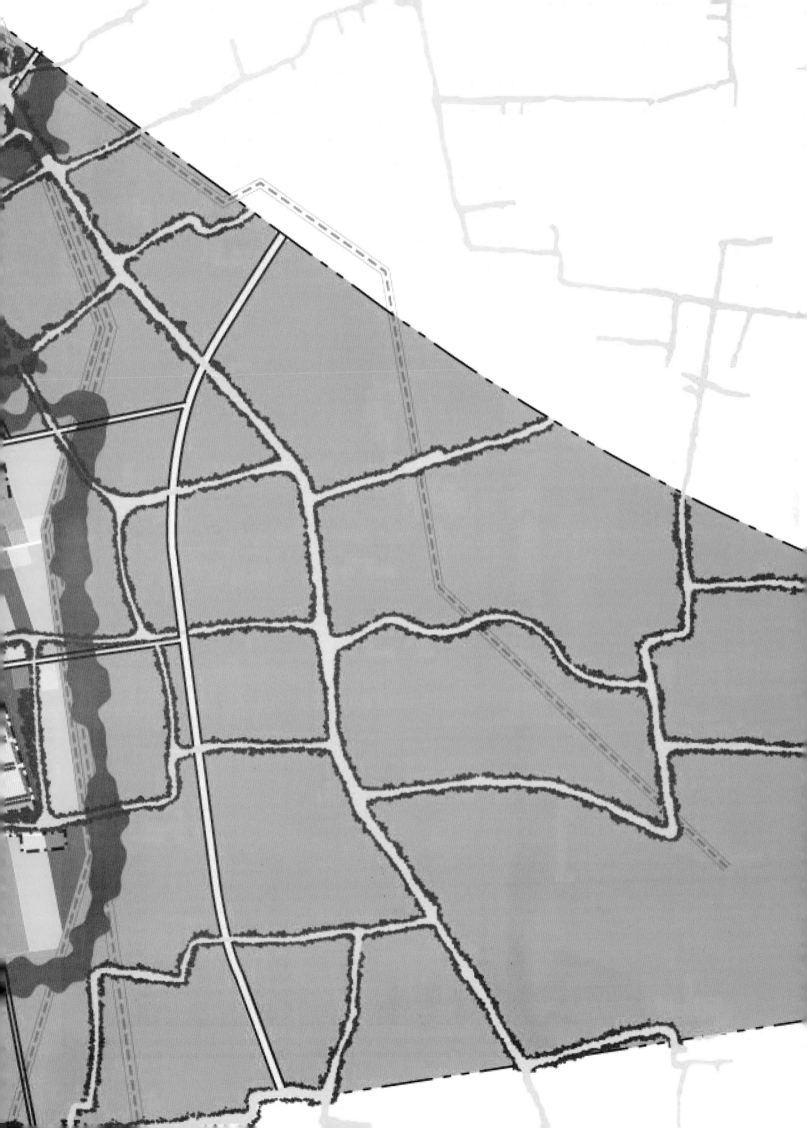

设计挑战

项目任务是设计一种新型开发模式，在保护现有耕地的基础上建设人口稠密的城镇村落。在近几年的案例中，即便是具备历史文化重要性的农耕土地也被铲平后让位于城市化建设。然而该设计需要充分利用基址中现存的农耕环境，借助交通运输与城市便利设施来建设村庄，使村庄结构紧凑、适宜步行、连接紧密，在保持农村风貌的同时使人们便捷地到达开放空间。这种保持基址现有农耕特色并与社区相融合的理念，真正地为中国当代农村生活提供了一种独一无二的营造方式。

在质疑由乡村农耕用地向城市建设用地流转的传统土地转型方式的同时，设计师们提出了结合新型城市发展的运营模式。这种模式通过引入规模经济效应、推广不同的开发密度，有效地提高了农地的产出。除此之外，通过净水湿地的营建和运河水网的重新组织，环境恶化问题也得以解决。

理想的解决方式

农田：在农业区的重新规划中，增加每家每户的耕种面积（普遍情况下，家庭农田面积为1~3 ha不等），开拓更大面积的优质高产区域（100 ha左右的有机农田）。在总面积为1100 ha的基址内，500 ha的土地将持续用作耕种农田，出产不同的农作物，如有机草药、蔬菜、水果、大米和鲜花。

人工净水湿地：经年累月，未经处理的农业用水和雨水径流直接排放进河系统，直接导致了现存运河的严重污染。为了缓解这个问题，项目设计引入一个45 ha的实体湿地净水系统来净化流经基址的污水。生物过滤器对污染物质进行吸收与消解，建造竣工的湿地将发挥从水中吸取富营养物质与重金属颗粒的生态功效。

现有基址内的运河系统包含很多断头的运河片断和一些互不相连的运河水道，这造成了河道堵塞、水质恶化。除了移除断头的运河之外，还需重新规整运河体系，以实现更好的流通和循环。完成河道的连接整理，就意味着物流人流循环交换系统的成功打造。在该系统中，运河河道将成为船只的水上交通通勤通道，运河滨河空间将会被用作人行步道。运河也将与村庄的生态雨洪管理和灌溉体系相结合。

滨水河岸网络：目前，基地内的河道网络中有很多尽端式河段且彼此没有连通的水路，因而造成水流停滞，水质恶化。项目方案将去除尽端式河段，重新连通河道体系，改善水流与循环条件。一旦完成系统连通，河道将用作水上交通航道，并沿河岸修建步道，形成一个流线网络。这一河道系统还将与新区内的生态雨水排放与净化系统相结合，具有雨水截流与生态过滤净水处理的功能，成为重要的景观基础设施，使水流先经过净化，再排入河道中。

雨水收集：基地内的道路、屋顶与其他不可渗透的表面积累的雨水径流将通过生态洼地进行收集、存储与净化。这类排水一部分将渗入地下，注入地下水层，另一部分将通过沿新区街道设置的生态洼地，导入河道或农田中。生态洼地中的植物将过滤并清除径流水中的污染物，再将径流水排入河道或农田中。

灌木篱墙廊道：在基地现有植被中，最突出的树种便是水杉。这种高耸的柱状乔木以紧密的南北向序列种植于基址之上可谓蔚为壮观。它们打破了农耕用地的大面积绵延，为私有耕地提供了隐私屏障，同样也串联起村庄与农田。贯穿南北的主路是以水杉作为行道树，呼应了基地内的本土农业树种传统。同样，水杉树林可引导基址上的主导风向，在炎热的夏季，为村庄带来丝丝凉爽。

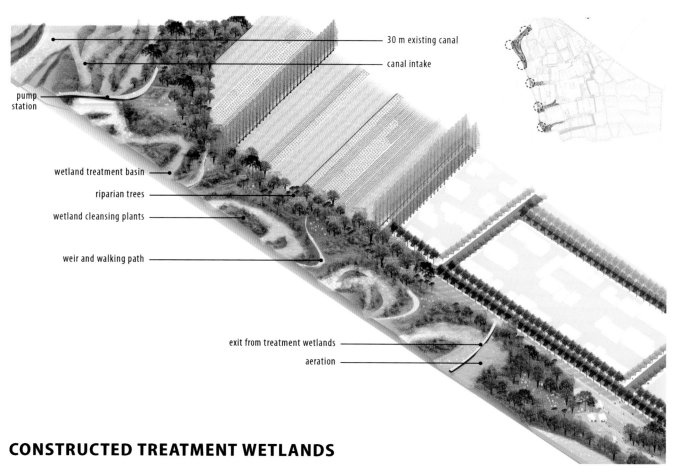

CONSTRUCTED TREATMENT WETLANDS

经年累月，未经处理的农业用水和雨水径流直接排入运河系统，直接导致了现存运河的严重污染。该项目引进一个45 ha的实体湿地净水系统来净化流经该基址的污水。

After years of untreated agricultural and storm water runoff discharging into the canal system, currently the existing canal water is highly polluted. This project utilizes 45-hectares of constructed treatment wetlands to cleanse the water before it enters the site.

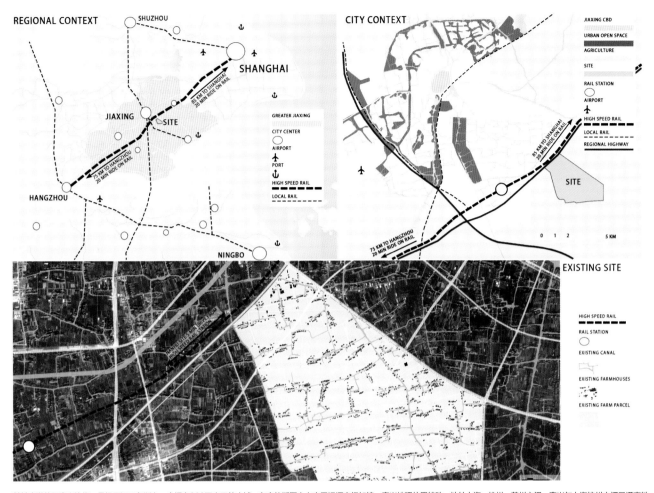

基址南湖位于嘉兴边缘，是扬子江三角洲上一个拥有300万人口的小城，在它的版图之上农田运河交织如锦。嘉兴地理位置特殊，地处上海、杭州、苏州之间。嘉兴与上海杭州之间已通高铁，车程约为20分钟。

Nanhu is currently a tapestry of small farms and canals on the edge of Jiaxing, a city of three million. Jiaxing is uniquely positioned between Shanghai, Hangzhou, and Suzhou, and is connected to Shanghai and Hangzhou by a 20 minute ride on the high speed rail.

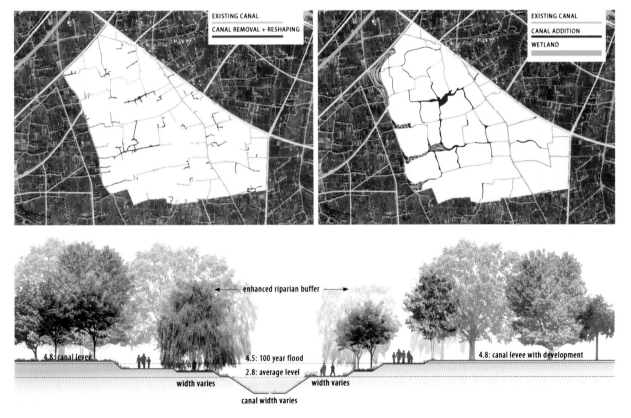

现有的运河系统中有很多互不相连的运河水道。除了移除断头运河之外，还需重新规整运河体系以实现更好的流通和循环。项目方案保留了现有运河的堤坝基础，为新运河设计了不同的景观堤坝。

The existing canal network includes many disconnected waterways. Dead-end canals will be removed, and the system will be reconnected to allow for better flow and circulation. Existing canals will retain their dike condition, and new canals will have a varying landscaped dike.

环境质量 + 可持续性

该项目的成功很大程度上得益于基址的环境质量,因此,可持续发展的建设模式尤为关键。建成的净水湿地可大幅净化水质。净化的运河系统也能使居民与游客直接参与其中。雨洪径流通过生态沟和雨水花园可回渗补给地下水,同时过滤污染物质。屋顶的雨水被收集用于农田的灌溉。水槽、盥洗、洗衣机产生的灰水被再利用为洁厕用水或非农业灌溉用水。有机农场中丰产的各种健康的当地农作物并不会对土地肥力造成过重的负担。在所有案例中,具有可持续性的解决措施和基础设施都是高度可观、可感、可实施的。

总结

随着快速发展与城镇化进程,未来中国的生存与环境质量很大程度上依赖于乡村农耕土地与城市建设用地之间的流转。工业化与城市的蔓延已经迅速吞噬了农田,中国需要一个范例,它用可持续性强并且意义深远的方式应对土地置换的过程。该项目将产量丰厚、适宜居住的密集型城镇村庄与现存的农耕土地有机融合,同时增加农业产量、改善环境质量。基于以上的原则实践,该项目为中国迫在眉睫的问题提供了答案。中国的城市化正在逐步深入,土地流转置换所衍生出的一系列问题不容忽视。通过展示现代化建设与融合发展的全新方式,该项目将在当代中国的未来发展中扮演愈发重要的角色。

Design Challenge

The project team was asked to design a new form of development that creates a dense urban village while at the same time retaining and enhancing the existing farmland. In many recent cases, agricultural land in China has been wiped clean of any previous historic or cultural significance to make way for urbanization. The design takes advantage of the site's agrarian setting and proximity to transit and urban amenities to create a village that is compact, walkable, and tightly knit, and yet has easy access to open space and maintains an agricultural character. This notion of retaining the existing agricultural character of the site and making it integral to the community provides a truly unique approach to contemporary village living in China.

The designers challenged the notion of typical rural-to-urban land transformation and created a model for integrated new city development. The design effectively increases farmland productivity by introducing economies of scale and promoting a variety of development densities. The design also addresses the existing degraded environmental conditions by introducing treatment wetlands and reorganizing the canal network.

Visionary Solutions

Farmland: The agricultural reorganization will increase the size of the individual farms, generally ranging 1-3 hectare size farms, create larger high

FARMLAND
The site's rich agricultural heritage is preserved and enhanced within the farmland.

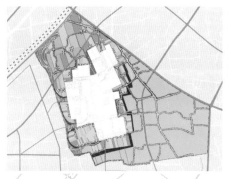

STORMWATER COLLECTION
Stormwater collection will be an integral part of the site's ecological water treatment and water cycle.

CONSTRUCTED TREATMENT WETLANDS
This project utilizes 45 ha of constructed treatment wetlands to cleanse the polluted water.

COMMUNITY PARKLANDS
Three major parklands with distinct character will have a regional draw and provide amenities for the district.

RIPARIAN NETWORK
The canal network will serve as a source of infrastructure, recreation, and circulation.

HEDGEROW CORRIDORS
Tall, columnar metasequoias will be planted in tight succession in north souths lines throughout the site.

基址内的这八处景观颇为关键。它们定义了基址并赋予它秉性,社区中的人们可以在此聚会休闲。它们呼应了现有的环境条件,在强化人们观景意图的同时也增强了景观的观感。

These eight landscapes are the most important moves on the site. They give the site identity and character, allow for community gathering and recreating, respond to the existing environmental conditions, and give visibility to the landscape intentions.

quality production areas, and a 100 hectare organic farm. Of the entire 1100 hectare site, 500 hectares will be maintained as working farms growing products such as organic herbs, vegetables, fruits, rice, and flowers.

Constructed Treatment Wetland: After years of untreated agricultural and stormwater runoff discharging into the canal system, the existing canals are extremely polluted. To mitigate this problem, the design plan utilizes a substantial 45-hectare constructed wetland treatment network to cleanse the water as it enters the site. Constructed wetlands perform the ecological function of extracting nutrients and heavy metals from the water by acting as a biofilter to absorb and break down contaminants.

Currently, onsite canal network includes many dead-end canal segments and disconnected waterways, causing stagnation and reduction in water quality. These dead-end canals will be removed, and the system will be reconnected to allow for better flow and circulation. Once connected, the canals will be used for water transport by boat and the edges and banks of the canals will have pedestrian pathways, creating a circulation network. The canals will be integral to the village's ecological storm drainage and treatment.

Riparian network: As it exists now, the onsite canal network includes many dead-end canal segments and disconnected waterways. This causes the water to stagnate and reduces the water quality. Dead-end canals will be removed, and the system will be reconnected to allow for better flow and circulation. Once connected, the canals will be used for water transport by boat and the edges and banks of the canals will have pedestrian pathways and serve as a circulation network. The canals will be integral to the village's ecological storm drainage and treatment. Stormwater run-off will be intercepted and treated by biofiltration landscape infrastructure before entering the canals.

Stormwater collection: Stormwater runoff from roads, roofs, and other impervious surfaces within the blocks will be collected, stored, and cleansed through bioswales. This runoff water will either permeate back into the water table or will make its way into the canals or agricultural fields via transportation by the bioswale along village streets. Plants within the bioswales will filter and extract pollutants from the run-off before they reach the canal or agricultural field.

Hedgerow corridors: The most prominent tree currently existing on the site is the metasequoia. These tall, columnar trees will be planted in tight succession in north south lines along the site for a stunning visual effect and to break up the large expanses of farmland, give privacy to individual

ECOLOGY PARKLAND

1. wetlands
2. wildflower fields
3. meadows
4. bird watching
5. marsh
6. pedestrian bridges
7. wildlife observation deck
8. pavilions
9. boating
10. boat house
11. picnics

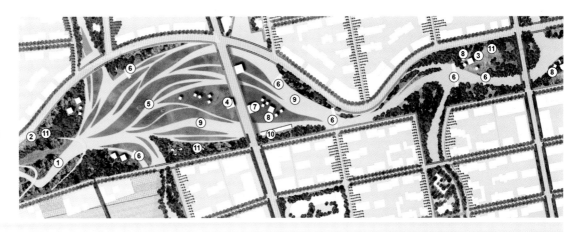

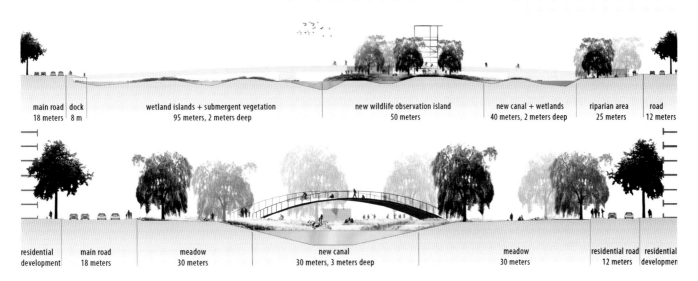

生态园地是净水湿地的延展，其中的绿地草坪、野生花境、长草沼泽是它的特色。野生动物观望梯地、鸟类观察点、庇护亭馆散落在生态公园之中。
The ecology parkland is an extension of the treatment wetlands, and features meadows, wildflower fields, and marshes. Wildlife viewing terraces, bird observatories, and shelter pavilions are placed within this ecological park.

farms, and link together the farmland and the village. Major connecting roads in the north-south direction will also be lined with metasequoias, adhering to the agricultural heritage of this tree to the site. These trees will also channel the prevailing breezes through the site, helping to cool the village during the warm summer months.

Environmental Quality + Sustainability

The success of the project depends heavily on the environmental quality of the site, and therefore sustainability issues are of primary importance. The constructed treatment wetlands will drastically improve the water quality. A cleaner canal system will allow residents and visitors to directly engage with the canal system. Stormwater run-off will infiltrate back into the groundwater via bioswales and raingardens, which will also work to filter out pollutants. Rainwater from rooftops will be collected for irrigation of agricultural land. Greywater from sinks, showers, and washing machines will be reused as a supply for toilets and non-agricultural irrigation. The vast parklands will help restore wildlife and habitat on the site. Organic farming with provide a plentiful source of healthy and local produce without treading heavily on the land. In all cases the sustainability solutions and infrastructure will be made highly visible, perceptible, and celebrated.

Conclusions

With its rapid development and urbanization, the future quality of life and environment in China depends heavily on the outcome of its rural to urban land conversion. As agricultural land is quickly eaten up for industrialization and urban growth, China needs a model to inform the process of land conversion with a sustainable and meaningful approach. This project answers that pressing question by illustrating the principles of how to integrate a productive and livable compact urban village with existing agricultural land while at the same time increase the productivity of the land and improving environmental quality. As China moves forward with ambitious growth and development, these questions of land transformation will only become more imperative. By demonstrating a fresh approach to modernization and integration, this project will be critically important in the future of contemporary China.

CONSTRUCTED TREATMENT WETLANDS

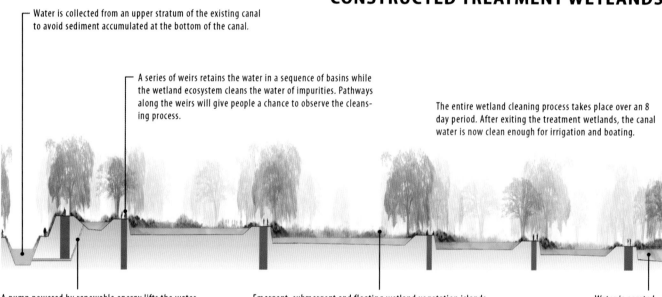

- Water is collected from an upper stratum of the existing canal to avoid sediment accumulated at the bottom of the canal.
- A series of weirs retains the water in a sequence of basins while the wetland ecosystem cleans the water of impurities. Pathways along the weirs will give people a chance to observe the cleansing process.
- The entire wetland cleaning process takes place over an 8 day period. After exiting the treatment wetlands, the canal water is now clean enough for irrigation and boating.
- A pump powered by renewable energy lifts the water from the canal and delivers it to the raised wetland cleaning system.
- Emergent, submergent and floating wetland vegetation islands metabolize the hydrocarbons, heavy metals, and other impurities from the canal water.
- Water is aerated upon exit.

除了发挥重要的生态功能，净水湿地也兼具教育启智与娱乐休闲的作用。堰坝上布置步道，以便人们漫步湿地之中，为其观看净化过滤工程提供机会。
Besides performing an important ecological function, the treatment wetlands will also be educational and recreational. Along the weirs there will be walkways inviting people to meander through the wetland area and giving a chance to observe the cleansing and filtration process.

STORMWATER CYCLE

Rain ....

Runnels: stormwater collection, stormwater transport ....

Bioswales: stormwater collection off of roads,
stormwater transport, and stormwater cleansing ....

Culverts: stormwater transport under roads ....

Remediation swale: stormwater cleansing, stormwater
transport to canals and agricultural fields ....

Clean water leaves the system by infiltrating back
into the ground or merging into the canals ....

雨洪径流通过生物沟得以收集、贮存与净化。径流的收集与生物沟明显地展示出净水处理过程是可观、可感、可供人体验的。

Stormwater run-off will be collected, stored, and cleansed through bioswales. The run-off collection and bioswales will be showcased and made evident so that people can experience the water cleansing treatment process.

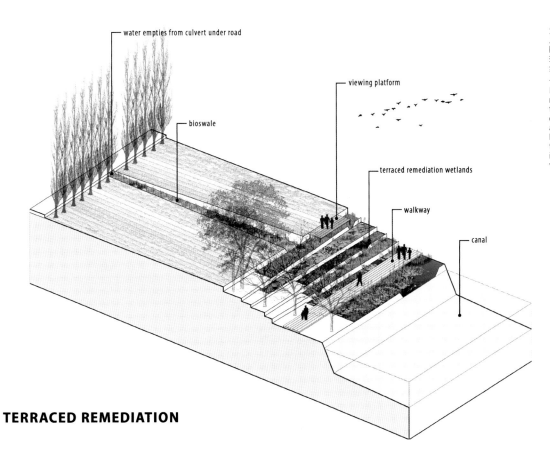

TERRACED REMEDIATION

- water empties from culvert under road
- bioswale
- viewing platform
- terraced remediation wetlands
- walkway
- canal

为防雨水径流需要额外处理，入渗沟会把径流运输至一系列阶梯式湿地盆地之中，径流在排空入运河前可以得到进一步净化，这些过程都是高度可见的。

In cases where the run-off requires extra treatment, the infiltration swales will deposit the run-off water into a series of terraced wetland basins that will further treat the water before emptying it into the canals, all in a highly visible manner.

114

奥芬巴赫港口区域规划
OFFENBACH HARBOUR PLANNING

项目位置：德国奥芬巴赫港
项目规模：25.6 ha+6 ha水域面积
设计公司：安博·戴水道
获奖信息：2011年DGNB可持续城市区域设计金奖

Location: Offenbach Harbour, Germany
Size: 25.6 ha+ 6 ha Catchaent Area
Design Firm: RAMBOLL STUDIO DREISEITL
Award: DGNB Sustainable City District Gold Award, 2011

美茵河上的一个工业半岛被转换成一个新型可持续城市区。"海港"形成新的城市焦点,通过多样的途径通往水边和多功能的城市空间。散步道网络系统一方面连接水体与密集的建成区域,另一方面串联新建地区公园和公园范围之外的地区河流道路。在土壤污染严重、充满挑战的环境中,一个全面的雨水管理概念被付诸实践,以营造柔质城市空间和街景,同时在雨水汇入河流和港湾之前,对它们进行储存和净化。净化群落等创新型天然水处理系统被整合入公园空间之中,新的自然栖息地被创造,为河岸边动、植物提供了栖息地,同时也为城市营造了一处清新的绿洲。该项目为在城市背景之下通过综合的工程手法营造富有吸引力的街区以适应气候变化提供了一个务实的解决方式。该项目被授予相当于LEED铂金奖的欧洲DGNB可持续性城区金奖。

An industrial peninsular on the River Main is being converted into a new sustainable city district. The "harbour" forms the new urban focus with varied access to the water and multifunctional urban space. A promenade network connects the water to the more densely build-up area on one hand and to the new local park and regional river path beyond. In the challenging environment of contaminated soil, a holistic stormwater management concept has been developed to create soft city spaces and streetscapes while retaining and cleansing stormwater before releasing it to the river and harbour. Innovative natural water treatment systems such as cleansing biotopes are integrated into the park spaces, and new natural habitats are created for riparian flora and fauna whilst proving refreshing green oasis for the city. This project is a pragmatic vision for how climate adaptation can be achieved in urban context creating attractive neighbourhoods through integrated engineering. The project has prequalified for the prestigious European DGNB Gold for a sustainable city district, comparable to a LEEDS Platinum rating.

富阳北支江滨江区景观规划

LANDSCAPE PLANNING OF FUYANG WEST RIVERFRONT

项目位置：中国浙江省富阳市
项目规模：90 ha
设计公司：SWA集团

Location: Fuyang, Zhejiang Province, China
Size: 90 ha
Design Firm: SWA Group

在中国传统文化中，"桓"象征着尊重、忠诚个人观点的表达以及永恒的爱情。该项目的设计灵感来源于民间故事和放鹿娃沉浸于自然山水美景的传说画面，设计师在项目场地中努力探寻山、水与当地文化中内含的"桓"元素。三个"桓"相互交错构筑于西部滨水大道与富春江之间的绿化带上。富阳西部滨江区是富阳市新兴的滨水开发空间，富春江沿岸新建了许多公寓住宅和大型综合建筑群。为了解决河道约10 m的水位落差，SWA集团开发了一种多层次滨水景观模式，既可有效防洪，也可对滨水地块进行灵活配置，赋予其多功能性。街道和高层建筑周边区域中注入了充满活力的景观元素，宜人的步道环境，在极富自然宜景的江畔地带，建有一座设施齐全的市政公园，园中景致与滨水美景相互融合。SWA对该项目8 km长的道路设施进行详细的分级处理，并对道路状况进行深入的研究，从而开辟出一条风景秀丽的滨水机动车道，且融入多样化的景观功能元素。全新的线形道路设计完好地保留了原有的茶园和相应的观景屏障设施，而街道景观设计则充分利用当地植物与多样化的空间元素配置，如墙体、遮阴树、多层次开花植物等进行合理布局。这些元素的应用体现了人行通道、十字路口、机动车道以及滨水漫步道之间鲜明的层次。

In traditional Chinese culture, "huan" represents respect, loyalty, expression of one's opinion, as well as everlasting, eternal love. The concept of this design is inspired by the folk story and imagery of the fictional Deer Boy enjoying the beauty of the mountain and the river, seeking and exploring within the site the huan of mountain, water and culture. The three huan intersect and overlap over the green belt between West Waterfront Boulevard and the Fuchun River. The Fuyang West Riverfront represents a new waterfront community that establishes apartments, condos, and mixed-use development along the Fuchun River in fast-growing Fuyang, China. To address the river's 10-meter fluctuation in water levels, SWA developed a tiered landscape that allows flooding and accommodates flexible, multi-purpose land use at the water's edge. Street-level and high-rise structures create a lively, pedestrian-friendly setting, with a formal civic park that blends into the river's naturalized edge. SWA's detailed grading and site studies for the project's eight-kilometer roadway integrates several landscape functions into one scenic riverfront drive. The new alignment preserves existing tea plantations, landscape screens, and views, while streetscape design makes use of a local plant palette and a variety of three-dimensional elements: walls, a formal framework of canopy trees, and layers of flowering accent plants. These elements express a strong hierarchy of gateways, intersections, and vehicular and pedestrian routes.

126

深圳华侨城欢乐海岸

OCT HAPPY COAST IN SHENZHEN

项目位置：中国广东省深圳市
项目规模：125 ha
设计公司：SWA集团，深圳市北林苑景观及建筑规划设计院
翻　　译：张旭

Location: Shenzhen, Guangdong Province, China
Size: 125 ha
Design Firm: SWA Group, Shenxhen BLY Landscape &Architecture Planning & Design Institute
Translated by Zhang Xu

欢乐海岸占地约125 ha。总体规划由两个既不同又互补的部分组成，即城市自然保护区与城市生活中心。每个部分根据两个不同的湖绕湖而设，其中一个是天然湖，另一个是人造湖。该人造湖建于天然湖与深圳湾之间的填海区域，使两片水域融于一体，把线性海滨公园与项目场地沿海岸线串联起来。因此，欢乐海岸总体规划的主导概念是"重新连接水体"。

景观设计师与客户和建筑师合作完成了该项目的总体规划。除了为所有水文要素、硬质景观、绿色屋顶以及种植区域提供全面的设计服务，设计师还引入了环境艺术的概念，聘请了一名艺术家进行艺术创作，并将作品协调地布置在园内。同时，景观设计师在深圳湾与两个湖之间引入了水循环系统，与水利工程师协作共同实现构想。

解决方案

该项目包括两个拥有不同项目元素的组成部分。

1. 城市自然保护区

该天然湖是在填海过程中形成的，湖的其中一边曾是深圳湾的海岸线。场地的设计主旨是围绕与水的互动，将水视为一种自然资源、美学元素、生态系统以及聚集设施。场地设计通过对沟渠进行填充与绿化，营造并扩展生态栖息地。场地的排水经过重新导向至滤沼泽地。过去用作海湾观察站的警戒塔，通过修复与增设亭台，现已改造为观鸟台。设计师通过扩建旧码头满足前往鸟类栖息地的船游需求，同时设置"租赁单车"供人们环湖骑游。所有铺垫材料都是循环再用的且具有渗透性，这符合可持续发展与最佳生态措施的目标。该自然公园拥有湿地、湖泊与林地约68万m^2，在密集的城市中为几十种野生动植物提供了栖息地，成为中国唯一地处城市腹地的滨海红树林湿地。

2. 城市生活中心

围绕中心湖区设置的城市生活中心，集文化、商业与娱乐等多元业态于一体，设有表演场馆、公共广场、公园空间与度假设施。水元素是设计的主导思想，并通过湖畔长廊、广场与公园得以体现。设计将水元素融入公共区域，贯穿项目主题。城市生活中心由七个分区组成，每个分区设置不同的主题，共同构成了一个整体。

社区影响与意义

项目的意义包括以下四点。

(1) 在一个缺乏公共开放空间的密集城市营造了一处以休闲与生态为中心的公益景观——新的中心公园。

(2) 作为满足经济要求的均衡发展范例，营造了开放的公共空间，保护并扩大了生态系统，提高了生活水平。项目还先后接待了许多中国政府官员与发展商，为其展示"欢乐海岸"的理念如何促进社区发展。

(3) 作为一个"活的博物馆"，为自然资源缺乏的城市提供了环境与生态教育。

(4) 整合的景观基础设施解决了泛洪与水质问题，同时提供了更多的休闲空间与相互连接的生态栖息地。

设计过程中面临的挑战与问题包括以下三个方面。

(1) 两个湖的湖水问题——一个自然湖，一个人工湖，其湖水都需要一个设计循环策略，以保持水质。除了引入周边城市区域的水之外，需要另外增加水源，以满足水循环的水量需求。海湾可提供一部分水源，但面临两大困难——潮位变化与盐碱化。经过景观师与水力工程师的通力合作，保持了两大湖的常水位，以促进海岸线设计与公众利用。景观证实，所增加的盐度均在恰当的范围之内，可满足湿地物种的生活需求，以及沼泽地进行过滤与提高城市地表径流需求。

(2) 由于项目需要大量的植被，设计师协助客户在项目现场施工的场地内设置了苗圃，满足了施工栽植前需要挑选植株和生长时期的要求。由于移植大型乔木的难度很大，这种实践性很强的方法至关重要。

(3) 景观设计师与建筑师及工程师合作，打造了一座独特的绿色屋顶建筑，在建筑设计中融入绿色屋顶技术与支撑体系。

区域鸟瞰图
Area aerial view

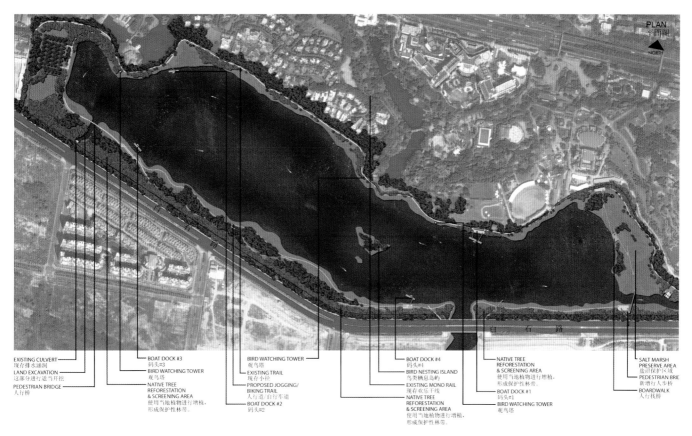

北湖湿地公园平面图
Site plan of North Lake Wetland Park

曲水湾鸟瞰图
Aerial view of Qushui Bay

The site area of the Happy Coast is proximately 125 hectares. The master plan is comprised of two different but complementary parts, namely the urban natural conservation area and urban civic center. Each of these two parts is based on a specific lake and established surrounded the lake, one of the two lakes is naturally formed and the other is manually constructed. The construct lake is located on the reclamation land between the natural land and Shenzhen Gulf, integrating the two water areas together, connecting the linear seaside park and the project site along the coastline as well. Therefore, reconnecting water bodies is the dominant concept of the Happy Coast master plan.

The master plan of the project was accomplished in the joint effort by the landscape architectures, the clients and the architectures. In addition to provide a comprehensive design service to cater the condition of the hydrological features, hard landscape, green roof and vegetation planting area. The idea of environmental art was adopted in the garden, the designers employed one artist to enrich the landscape, the artworks created by the artist were placed harmoniously on the garden site. Meanwhile, the landscape architecture invited the concept of water circulation between Shenzhen Gulf and the two lakes, and collaborated with hydraulic engineer to achieve this envision.

Solution Plan

The project is comprised of two components of which consist different programmed features.

1. Urban natural conservation

The natural lake is formed in the process of sea reclamation. One bankside of the lake was the former Shenzhen Gulf coastline. The design purports to involve the idea of water interaction, perceiving water as a kind of natural resources, aesthetic element, ecological system and gathering facilities. By refilling and greening the channel, the designer succeeded to creating and enlarging the ecological habitat on the site. The reorganized drainage system conveyed the water to the filtration wetland. The former guard tower once used as the gulf observation station now transformed as a bird-seeing port, by the means of restoring and increasing the number of pavilion. The designers expanded the old pier to improve the ferry service in order to cater the visitor's bird habitat visiting requirement, meanwhile, establishing the bicycle renting port to facilitate the visitors' round lake sightseeing. All the padding materials are both recyclable and permeable. This is accord with projective of sustainable development and optimal ecological measures. This natural park contains wetland, lake, forest, with a total area of 680 000 square meters, provide natural habitat to tens of species of wildlife in the compact city, and became the only mangrove forest wetland which located in central urban area of China.

2. Urban life center

The central lake area was surrounded by the multi-model urban life center with arena, public plaza, park space and holiday facilities, which links the function of culture, business and recreation together. Water, as the main element in the whole design process, interpreted by the corridors along the lakeside, squares and parks. The hydraulic factors is integrated in the public area, and permeated in the keynote of the design. The urban life center consists of seven zones. Each zone has a distinct theme, and is an organic part of the whole park.

Community Effect and Meaning

The meaning of the project could be concluded as follows.

(1) Creating a public welfare landscape dedicated in recreation and ecology in the compact urban area, where is severely lacking of public open space. It is the new central park.

(2) It's a model for meeting the balanced requirement of economy and development, creating public open space, protecting and enlarging the ecological system, and improving the citizens living standard. The project received a series of government officials and developers for a visiting, demonstrating how the Happy Coast's design envision devoted in helping the development of the community.

(3) As a living museum, it provides environmental and ecological education for the natural resource deficient city.

(4) The integrated landscape infrastructures address the issues of food and water quality. Meanwhile, they provide more mixed leisure space and ecological habitat.

Challenge and problems the project has team faced and tackled during the design process are as follows.

(1) The issue of water quality in the two lakes: one is a natural lake and the other is a constructed lake, a designed circulation strategy should be employed to guarantee the water quality and purification. In addition to introduce the water from the other adjacent urban area, but also increase the water sources to ensure the circulated water quantity. The gulf can be an additional water source, but great difficult still existed, namely tidal fluctuation and salinization. Owing to the cooperation between landscape architectures and hydraulic engineers, the water elevation stabilized in a consistent level. This facilities and enhances the design and public use of the coastline. The effect of landscape verified the fact that the slightly increased salinity within an appropriate range, not only can inhabit the wetland species, but also helped the wetland filtration process and increased the city ground surface runoff requirement.

(2) Since abundant of plants were required in the project, two years ago, we assisted our clients to establish the nursery garden in advance of the construction work. This move meets the needs of plants selection and growth period. This measure is feasible and practical, and would avoid the difficult happened in the big tree transplanting practice.

(3) A unique green roof building was constructed during the collaboration with architects. The landscape architects proposed a green roof technique and support system in the architecture design.

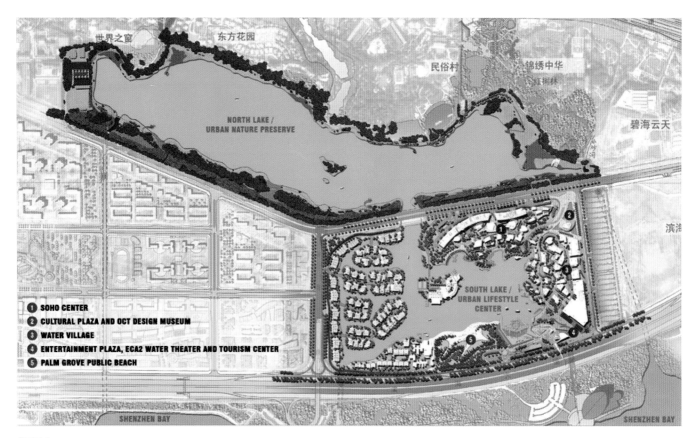

总平面图
Site plan

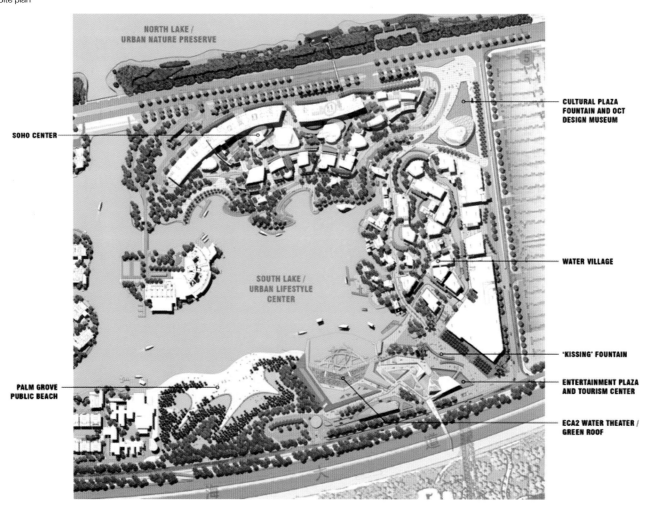

商业区平面图
Business district plan

134
深圳北站商务中心区城市绿谷景观规划设计
URBAN GREEN VALLEY LANDSCAPE PLANNING & DESIGN IN NORTH STATION CENTRAL BUSINESS DISTRICT, SHENZHEN

项目位置：中国广东省深圳市
项目规模：125 ha
设计公司：安博·戴水道，深圳蕾奥城市规划设计联合体
设计时间：2015年4月
翻　　译：张旭

Location: Shenzhen, Guangdong Province, China
Size: 125 ha
Design Firm: RAMBOLL STUDIO DREISEITL，LAYOUT
Translated by Zhang Xu

深圳北站商务中心区是深圳市委市政府确定的13个重点区域之一,为高效落实市委市政府关于加快推进全市13个重点区域开发建设的统一部署,高水平打造深圳北站商务中心区,实施"一核、一心、一网"重点片区建设计划。"一网"即城市绿谷,贯穿整个北站周边地区,连接羊台山生态资源环境与城市地区的生态绿色网络。其不仅是展示新区自然景观风貌的直接载体,也是带动地区发展的重要引擎;其不仅是承载城市公共活动、传承城市文脉的核心场所,更是推动片区可持续发展的基础,对北站地区城市环境的高品质塑造有着深远的影响。

该设计采用"EOD+TOD"的设计理念,将绿谷作为城市绿色公共空间,以"生态绿谷、活脉融城"的姿态,打造革命性的绿廊,从自然、人和城市三个方面着手,以9km动感纽带为核心组织总体景观框架,具体设计包括以下几个方面。

(1)通过建立绿色可持续的生态网络,塑造自然乡土景观,开展可再生能源和园林垃圾的循环利用,构建低碳便民的慢行网络,实现从破碎失衡的生态环境到海绵城市的转变。

(2)通过打造人本生活圈,融合多元感官体验和多彩空间,经由不同高度、流畅观赏体验的立体动线和街道,将激情与活力延展开来,实现从单调的社区到人文活力乐园的转变。

(3)通过塑造以绿色生态、缤纷多元的9km动感纽带,展现创新创意、绿色健康的六大城市庆典和精彩纷呈的九大缤纷景致,实现从平淡乏味的印象到缤纷魅力名片的转变;通过协调时序、积极互动的发展,推动城市实力的整合提升。

后续,安博·戴水道将与其他8家设计机构对项目设计进行深入讨论与整合,以完成最终成果的编制与展示。

总平面图
Site plan

人文活力乐园：乐活之心 + 动感空间

人本生活

设计理念：人本生活圈

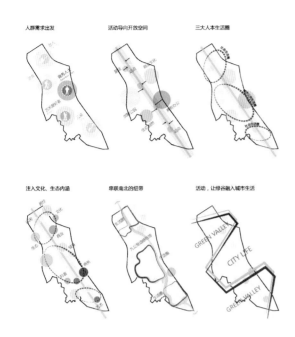

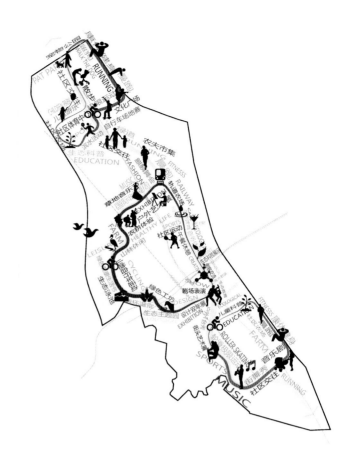

缤纷魅力名片：魅力形象 + 互动发展

一带六节九景

绿谷以绿色生态、缤纷多元的**九公里动感纽带**，展现创新创意、绿色健康的**六大城市庆典**和精彩纷呈的**九大缤纷景致**塑造独具魅力的城市名片。

一带——九公里动感纽带

利用绿谷不同区段的本底特征，营造主题多元的绿谷公园群，塑造独特体验。

纵深联系的场所，塑造连续性、复合型的景观路径，把人引到山水边，引到城市中心，让人文体验贯穿其间，把城市精神和活力传递到健地去，成为龙华新城的核心名片。

六节——六大节庆活动

引入六大节庆活动，包括龙华·绿谷绿色生活节、龙华·绿谷健康生活节、深港设计双年展会、深圳国际时装周、深圳国际创客周，在丰富市民日常生活的同时，通过承的国际化的活动来提升城市知名度，塑造城市的品牌形象。

九景——九个景观核心

着重打造绿谷九大景观核心，包括以绿色健康生活为特色的上塘社区公园景观核心，以文化休闲体验为特色的城市绿谷公园景观核心，以文化庆典体验为特色的红山站公园景观核心，以生态郊野景观为特色的红木山郊野公园景观核心，以生态果科展示为特色的北站公园景观核心，以生态果科主题体验为特色的城市公共庆典为特色的南园公园景观核心，以极限运动为特色的民丰玉龙公园景观核心，以及以音乐主题为特色的白石龙公园景观核心。

Shenzhen North Station Central Business District (CBD) is one of the 13 key areas granted by the Shenzhen municipal government. In order to efficiently implement the integrated-deployment strategy of accelerating the development and construction in the 13 key areas in Shenzhen, a high-standard North Station in Shenzhen Central Business District should be established, and a major area adopting the One Core, One Heart, One Network idea should be constructed. One Network refers to the city Green Valley. It is a system weaving through the whole area adjacent to the North Station, and linking Yangtaishan ecological resources & environment with the ecological green network in urban areas. It is not only demonstrates the natural landscape view in the new area directly, but also efficiently propels the regional development. It can be a brilliant platform for the urban public activity, a vital site to inherit the civic culture, and a basement for the sustainable development in this area. All these have a profound influence on shaping and creating the high-quality urban environment in the North Station area.

The design team employs a EOD + TOD design concept, treating Green Valley as the urban green public space, in the vision of permeating the city into the green valley and recirculating the city through the green valley, they created a set of revolutionary green corridors. Besides, based on three aspects, namely nature, human and city, they framed the nine kilometers dynamic link as the core area of the plan, and organize the overall landscape along them. The design details include several parts.

(1) Through the establishaent of green and sustainable ecological networks, create and reshape the local natural landscape, develop and utilize the renewable energy, recycle and reuse the garden waste, forge a low-carbon consume pedestrian network for the convenience of people, achieve the transition from an imbalanced fragmental habitat to an integral ecological sponge city.

(2) Through the creation of a human-oriented living circle, integrate the multi-sensory experiences into colorful spaces, regenerate the passion and vitality through seamless sightseeing experience provided by the tree-dimensional activity-routine and the streets with different elevation, realize the conversion from a monotonous and tedious lifestyle to a vigorous humanistic park.

(3) Through creating the green ecological and multi-featured nine-kilometer dynamic link, interpret the 6 city's celebration with innovative & creative notion and green and health idea, demonstrate the 9 types of urban landscape view with diversity & colorfulness. Therefore, a doll and boring city impression was substituted in an attractive and distinctive unban charm and charisma. Coordinate the time-sequence and positive interaction of development, finally accomplished the objective of city integration and promotion.

To follow up, the final compiling work of the task would be eventually finished during the further and tighter collaboration between Atelier Dreiseitl and the other 8 design institutions.

生态海绵城市：EOD 自然生长 + TOD 绿色出行

海绵城市

海绵城市理念

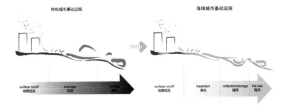

海绵城市水管理战略

传统的城市水管理模式中，雨水径流和没有处理的污水直接排入河道，河道将以最快的排洪速度将污染的水排入海洋，从而造成水污染和生态退化、内涝以及淡水资源浪费的系列问题。我们提出针对龙华新区水管理的海绵城市策略：将传统的风险管理转变为分散式的水管理模式，强调用水资源在城市中的导流、源头净化、储存、缓慢排放至下游，融入城市的环境和生态网络中。

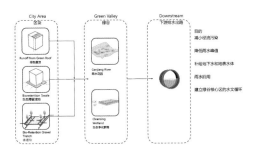

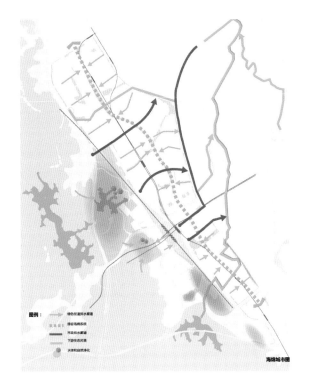

海绵城市图

生态海绵城市：EOD 自然生长 + TOD 绿色出行

永续利用

能源资源问题

随着城市化程度的进一步提高，绿谷和周边城市区域采用常规能源及处理方式将会增大电量消耗、市政资金投入、增加垃圾填埋场负担，甚至引起污染问题。

绿谷解决途径

作为城市低碳发展的重要载体，绿谷将通过可再生能源和园林有机垃圾循环利用来降低绿谷环境运营的依赖。如果将绿谷中的园林绿化垃圾就地源头处理，不仅可以变废为宝，循环利用，也将成为深圳城市可持续发展的样板示范工程推广到更多的区域应用。

47800 吨/年
可再生园林垃圾

绿谷绿地 239 公顷
园林垃圾年产量 47800 吨

1.3×10¹⁰ MJ/年
太阳能辐射

北站地区属于太阳能资源丰富区域，对土地资源极为紧张的深圳来说，规模化的太阳能发电站可行性不大。绿谷提供了将太阳能分散式利用的可能，通过光伏材料和景观设施的结合，打造互动性强的新能源绿谷景观。

有机土壤循环展示中心   有机园林土壤

永续利用

永续利用是景区景观和基础设施系统设计、建设运营过程中的一个重要部分。绿谷系统承担低碳建设的功能，打造成为与周边社区基础设施连接的可持续联网公园。通过构建太阳能景观、园林垃圾循环利用，结合自行车快线系统等，达到降低能源依赖、减少温室气体排放、减少废气物、低碳出行等方面的总体发展目标。绿谷将成为一个正能量的公园，降低城市对灰色基础设施的需求，保证城市的可持续发展。

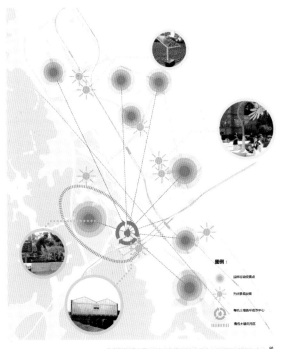

142

宁波生态廊道

Ningbo Eco–Corridor

项目位置：中国浙江省宁波市
项目规模：600 ha
设计公司：SWA集团
翻　　译：张旭

Location: Ningbo, Zhejiang Province, China
Size: 600 ha
Design Firm: SWA Group
Translated by Zhang Xu

通过对地形、水文和植被的创新性融合，宁波生态廊道项目使原本丧失栖息地供给性的棕地蜕变成可增强生态系统丰富性与多样性的活态过滤器。这条3.3 km长的生态廊道，不仅促进了人类活动与生物栖息之间的协同增效作用，同样也是在经济迅猛发展的语境下，中国城市扩张与发展中一个弥足珍贵的教育启智范例。

背景

宁波位于中国东部海线的扬子江三角洲腹地，城市人口高达349万人（2010年人口普查数据）。宁波是中国历史悠久、声名远扬的城市之一，它不仅是关键性海外贸易通商口岸，同时也是举足轻重的经济中心。正如中国其他城市一样，近些年来显著的人口增长，不仅给宁波的基础设施建设施加了巨大压力，同样，统筹协调城市密集稠化与降低负面生态影响也给政府带来了前所未有的挑战。

在2002年，为了缓解老城区压力，并为城市无序蔓延树立一个兼顾平衡与生态的手法先例，宁波规委提出了建设"宁波东部新城"的总体规划。该项规划围绕"生态廊道"组织6 km²的多用途导向城市开发，主要由线状绿色空间体系组成。在该体系中，人类、野生动物与植物能够栖息、共存与繁荣。

生态区域语境

宁波位于长江平原南部，地处长江（扬子江）三角洲的低腹地段，在生态区划上属于常绿阔叶林带。该地区在历史上以广袤的常绿橡树林和长草沼泽环绕的季节性洪泛湖泊盆地地貌著称。持续百年的农业耕作与近期以来的城市发展导致湿地的大量干涸与水域栖息地的巨幅减少。仅存的湿地成为东方白鹳、天鹅、西伯利亚鹤与白颈鹤等候鸟的重要栖息地，也成为扬子豚、扬子鳄、中华水鹿和水獭等水栖动物的家园。

这些湿地与水域栖息地能对这片生态区域起到至关重要的保护作用，设计团队基于此而将主要精力投放在湿地恢复上，并以此作为一种基址专项介入（干扰），以期与这片具有生态自我修复性的新区域建立历史联系并实现文化传承。

场地调研

宁波地区因其运河系统而享有盛誉，在历史上，运河兼具防洪、灌溉与运输功效。在该生态廊道基址内，在有效区划与污染控制缺失的情况下引入工业生产，导致运河功能与形态的严重退化。随着工厂的迭代建设，区域内的建设作业中挖掘出的污染性土方被非法堆放随意丢置。运河被用作临时排放填充沟渠，工业废水和雨洪径流直接排入其中，致使运河堵塞水流滞缓。若使人工介入效果卓然、意义深远，则需采集可充分反应基址状况的数据。因此，领队景观设计师和协同人员，包括水质科学家、湿地专家、水文工程师，对场地进行了深入的分析。为了全面了解基址现存状况，设计团队绘制了当地水文圈层图和基址自然水流图，并且甄别出潜在的协同作业可能性。

实施：建立活态过滤器

很多水道假途地势低洼、地貌蜿蜒的山丘，将它们规整地组合成网络系统，以治理已建成运河系统内的污水，处理新开发区域的雨水径流，构建滨河区域来恢复野生栖息，并为新迁居民提供娱乐休闲与教育启智的机会。

水文：新塑弯曲水道提升水文状况

现有的运河系统中，有的河网已成断头死水，有的河网已经支离破碎，这将被一系列自由流淌的涓流、小溪、池塘和沼泽所代替。为了支持本地生态的重建，设计团队较好地控制了水文流态，使之趋于流速缓慢、路径蜿蜒，并接近蓄洪平原低地的原始状态。

1	Water Jet
2	Expanded Water Body
3	Wind Mills on Mound
4	Outdoor Teaching Space
5	Nature Study
6	Underground Garbage Facility
7	School
8	BBQ Area
9	Land Art Pedestrian Bridge
10	Bio-Retention Basin
11	Bio-Ponds
12	Wellness Gardens
13	Sand Volleyball Court
14	Children's Playground
15	Bio-Dry Creek
16	Main Pedestrian Loop with Bike Lane
17	High Rise Residential
18	Pedestrian Bridge over Main Creek
19	Pump House Facility
20	Outdoor Swimming Pool
21	Sculpture Garden
22	Campus
23	Water Cleansing System
24	Pedestrian Bridge and Overlook
25	Boat Dock
26	Observation Tower
27	Children's Learning Center
28	Basketball Court
29	Skateboard Park
30	Volleyball Court
31	Parking Lot
32	Community Village
33	Water Front Platform
34	Rock Climbing Area
35	Neighborhood Center
36	Nature Study
37	Off Stream Wetland
38	Community Garden
39	Boardwalk
40	Water Edge Promenade
41	Primary Wetland

除了通过模仿当地生态过程而使用创新的生态修复技术之外,新建的水道也将现存的运河V级水质(工业、农业用水限制使用,居民生活用水不宜使用)提升至III级水质(生态修复用水与娱乐用水适宜使用)。

地形:山丘与峡谷体系引导水流方向

将周边开发区域的土地挖方与填方相结合,整个生态廊道区域经过精确评价测算之后赋予不同高差而塑形成山丘峡谷体系。这种峡谷水道通过沉积作用、曝氧作用、生物处理作用移除污染物质,通过滞留作用补给蓄水层水分,并且随着水流经基址,水道的形貌也进一步加强。除此之外,山丘系统可以用作城市环境的生态界限,为新建城市构建景观风貌,为游人提供观景延展点,并增强栖息地的多样性。

植被:当地植被清洁水源、创造栖息条件

基于审美性、纲领性、生态性与气候性的考量,策略性地更替落叶树种与常绿树种。在廊道长度范围内重视乡土物种的使用,支撑并强化当地植物群落多样性的重建,和本土野生动物组群的繁殖。基址上的滨水植被、生态沟和雨水花园可清洁周边发展区域、建设与硬化用地的雨水径流。植被的甄选也彰显出独一无二的场所感:协同地形的多样性,根据标高、肌理、色调的不同区分使用物种,进而形成组群来创造独特的空间模式。

生态廊道是宁波新城开放空间系统的脊梁,绵延3.3 km,创造并衔接了一系列的土地效能,绵延3.3 km。生态廊道无缝的缝合在周边的城市肌理与自然系统中,构建了绿道与周边景观环境标志性的关系。

通过恢复本地区的生态系统,宁波生态廊道为当地植物区系与动物区系营造了重要的栖息环境,为当地与周边社区提供了惬意宜人的公共空间,同样也为中国和其他地域的可持续发展工作添了砖、增了瓦。

Through the innovative synthesis of topography, hydrology and vegetation, the Ningbo Eco-Corridor project transforms an uninhabitable brownfield into a 3.3 km long "living filter" designed to restore a rich and diverse ecosystem, create synergy between human activity and wildlife habitat, and serve as valuable teaching tool and model for sustainable urban expansion and development in China's rapidly advancing economy.

Background

Situated on China's eastern coastline in the heart of the Yangtze River Delta, with an urban population of 3.49 million (2010 population census), the city of Ningbo is one of China's oldest and best-known cities, a key port for foreign trade, and an important economic center. As with many other cities across the country, phenomenal population growth in recent years has put an enormous strain on infrastructure, posing a monumental challenge for the local government to accommodate urban densification while minimizing negative environmental impacts.

In 2002, with the intention of alleviating pressure on the Old City while setting a precedent for a balanced, ecological approach to urban expansion, the Ningbo Planning department called for a master plan for the creation of the "Ningbo Eastern New City." The plan would include 6 square miles of mixed-use urban development organized around a signature "Eco-Corridor," comprised of a linear network of green spaces in which humans, wildlife and plants could inhabit, co-exist, and thrive.

Eco-Regional Context

Ningbo is situated in the southern part of Changjiang Plain Evergreen Forest Eco-Region within the low-lying Changjiang (Yangtze) River delta. This eco-region was historically characterized by extensive evergreen oak forests and reed swamps surrounding seasonally-inundated lake basins. Centuries of agriculture and recent urban development have resulted in a

EXISTING ISSUES

DEGRADED WATER QUALITY

stagnic water

pulluted discharge from factory and farm

DEGRADED HABITAT

wildlife endangered

riparian buffer replaced by hard edge

ECO-CORRIDOR AND LANDUSE PLAN

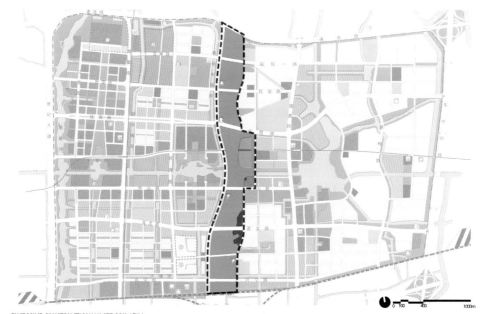

EXCESSIVE CONSTRUCTION WASTE SOIL/ FILL

significant loss of wetland and aquatic habitat. The few wetlands that remain provide vital habitat for migratory birds such as oriental white storks, swan geese, and Siberian and white-naped cranes, and aquatic animals such as Yangtze River dolphins, Yangtze alligators, Chinese water deer, and otters.

Recognizing that wetland and aquatic habitats hold the greatest conservation significance for this eco-region, the design team focused its energies on wetland restoration as the impetus for a site-specific intervention that would have historical and cultural relevance in this new era of eco-consciousness.

Site Investigation

The Ningbo region is characterized by a canal system that historically performed a set of functions including flood control, irrigation and transportation. Within the Eco-Corridor site, the canals had become severely degraded with the introduction of industrial uses in the absence of effective zoning and pollution controls.

As successive generations of factories were built, contaminated soil from construction excavation was dumped illegally and randomly through out the area, with factory sewage and stormwater runoff allowed to flow untreated into the canals rendered stagnant by ad-hoc infill.

Knowing that an effective and meaningful intervention would require sufficient data about the underlying conditions of the site, a thorough analysis was conducted by the leading landscape architects and affiliated consultants—including water-quality scientists, wetland experts, and hydrologic engineers—in order to fully understand the existing conditions, map the local hydrological cycle and natural flow of water across the site, and identify potential synergies.

Implementation : Building A Living Filter

Out of this analysis emerged the concept of creating a microcosm of the Changjiang Eco-region. A networked series of waterways organized by low, undulating hills was designed to treat polluted water from the established canal system, manage stormwater runoff from the newly developed areas, establish riparian zones for the restoration of wildlife habitat, and provide recreational and educational opportunities for the new inhabitants.

Hydrology: a new meandering watercourse to improve hydrological function

Replacing the existing system of dead-end and disconnected canals is a series of free-flowing rivulets, streams, ponds and marshland. The hydrological flow is designed to be slow and meandering, approximating the original conditions of the lowland floodplain, in order to support the re-establishaent of the native ecology.

Through innovative bio-remediation technologies that mimic indigenous ecological processes, the newly constructed watercourse improves the existing quality of the canal water from Class V, restricted to industrial and agricultural uses and not fit for human habitation, to Class III, which is suitable for ecological restoration and recreational use.

Topography: A System of Hills and Valleys Directs Water Flow

Incorporating fill from excavation in the surrounding development areas, the entire Eco-Corridor zone is carefully graded and shaped into contours creating a terrain of hills and valleys. The valley waterways serve to remove

BUILDING A "LIVING FILTER"
Microcosm of the eco-region

HYDROLOGY : a new meandering watercourse to improve hydrological function

- Existing dead-end and disconnected canals
- Proposed connections and water bodies
- Meandering water system

Bio retention pond　　　Off-stream wetland　　　Marsh wetland

Hydrology filter layer: Replacing the existing system of dead-end and disconnected canals is a series of free-flowing rivulets, streams, ponds and marshland that will support the re-establishment of the indigenous ecology.

pollutants through settlement, aeration and bio-processing, allow retention for aquifer recharge, and highlight the different modalities of water as it moves across the site. The hills also serve to buffer the urban environment, frame views to the New City, provide vista points for visitors, and increase habitat diversity.

Vegetation: Native Plantings Cleanse Water and Create Habitat

Across this undulating landscape, the strategic placement of deciduous and evergreen species reflects aesthetic, programmatic, ecological, and climatic considerations. An emphasis on native vegetation supports the re-establishaent of diverse plant communities along the length of the corridor, and encourages colonization by indigenous wildlife. Plantings along the riparian edge, and bio-swales and rain gardens throughout the site, cleanse stormwater run-off from the adjacent development and other building and hardscape areas. Plant selection also creates a unique sense of place: together with topographical variety, differentiation of species into groupings based on height, texture and color creates distinct spatial patterns.

The Eco-Corridor serves as the spine of Ningbo New City's open space system, creating and connecting a variety of land uses. Extending 3.3 km, the Corridor merges seamlessly with the adjacent urban fabric and natural systems, creating a symbiotic relationship between the greenway and surrounding landscape.

By restoring the ecological network in this region, the Ningbo Eco-Corridor creates vital habitat for native flora and fauna, enhances public health, creates fun and enjoyable public spaces for local and neighboring communities, and raises the bar for sustainable development in China and beyond.

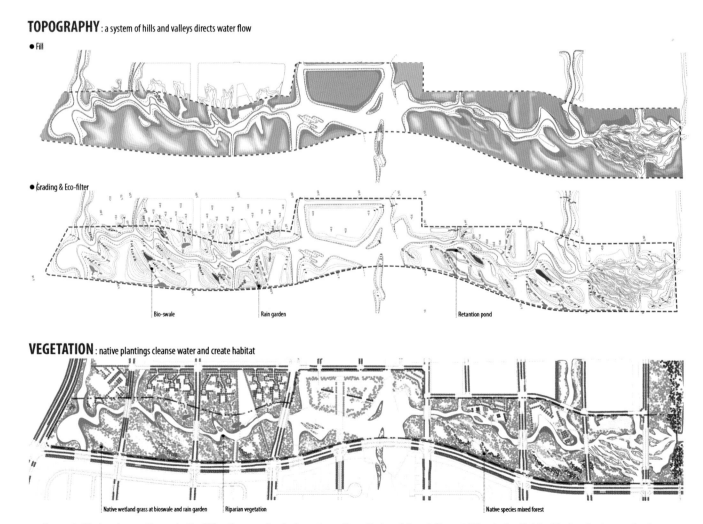

Topography filter layer: Incorporating construction fill from the surrounding development areas, the corridor is carefully graded to create hills and valleys that direct the flow of water across the site.

Vegetation filter layer: Native vegetation supports plant diversity and encourages colonization by indigenous wildlife. Variegated plantings also serve to purify ground water, cleanse stormwater run-off and create a unique sense of place.

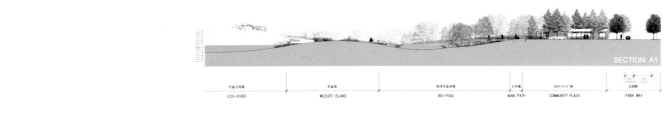

SECTION: A1

| ECO-RIVER | WILDLIFE ISLAND | BIO-POOL | MAIN PATH | COMMUNITY PLAZA | PARK WAY |

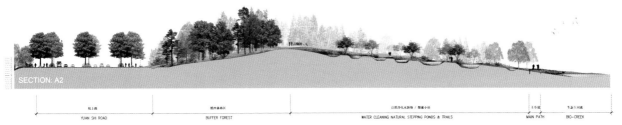

SECTION: A2

| YUAN SHI ROAD | BUFFER FOREST | WATER CLEANING NATURAL STEPPING PONDS & TRAILS | MAIN PATH | BIO-CREEK |

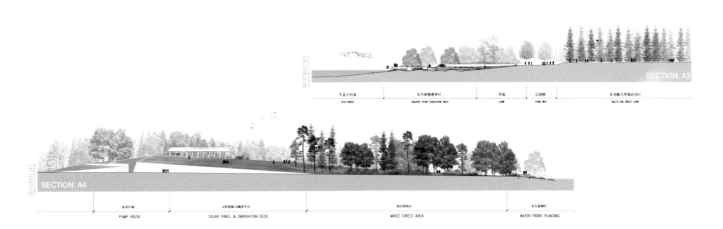

SECTION: A3

| ECO-CREEK | AQUATIC PLANT EDUCATION AREA | LAWN | PARK WAY | MULTI-USE GREAT LAWN |

SECTION: A4

| PUMP HOUSE | SOLAR PANEL & OBSERVATION DECK | MIXED FOREST AREA | WATER FRONT PLANTING |

WATER QUALITY IMPROVEMENT

● WATER TESTING & SAMPLING

PHASE 1 PLAN

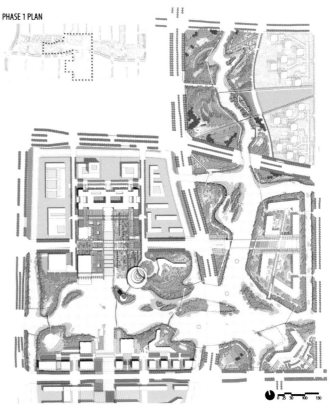

● GOAL : IMPROVE WATER QUALITY FROM CLASS V TO CLASS III

EXISTING	CLASS V	CLASS III	UNITS
DO	2.48	5.0	PPM
BOD	7.16	4.0	MG/L
COD	38.00	15.0	MG/L
TSS	20.00	15.0	MG/L
NH3	5.53	1.0	MG/L

● STRATEGIES

1. increase waterflow. ········→ ACTIVE WATER TREATMENT
2. remove targeted pollutants. ········→ WETLAND
3. harvest clean rain water. ········→ STORMWATER MANAGEMENT

Water sampling and mapping of the local hydrological cycle and natural flow of water across the site inform strategies for water quality improvement from Class V, restricted use to Class III, suitable for recreational activities.

ACTIVE WATER TREATMENT: increase waterflow

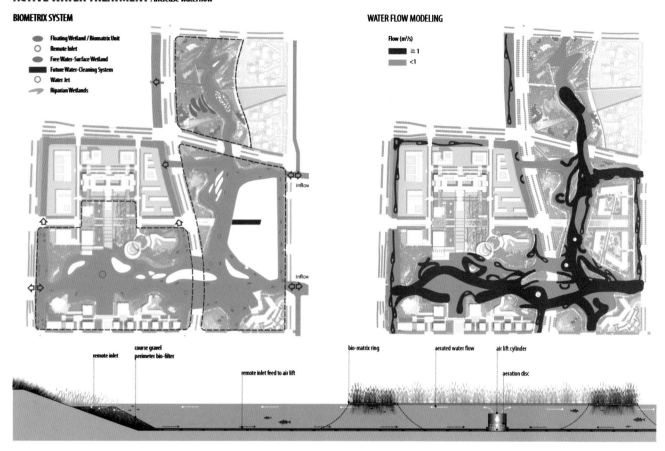

In collaboration with water-quality scientists, a system is designed using passive and active methods to aerate and encourage water movement over plant roots to remove contaminants.

WETLAND: remove targeted pollutants

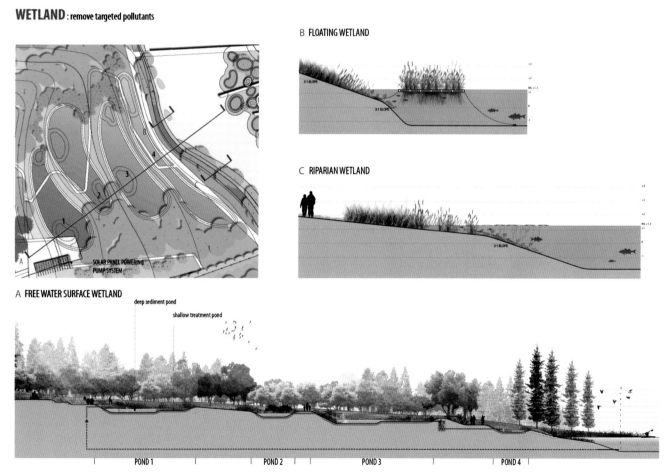

In collaboration with wetland experts, a site-specific system of free water surface, floating, and riparian wetland are designed to remove targeted pollutants.

STORMWATER MANAGEMENT : harvest clean water back to ground and stream

- Grass Land & Wood Land
- Green Building
- Permeable Paving
- Bio-Swale
- Rain Garden
- Flood Storage
- Run-off Treatment
- Roadway Run-off Filter

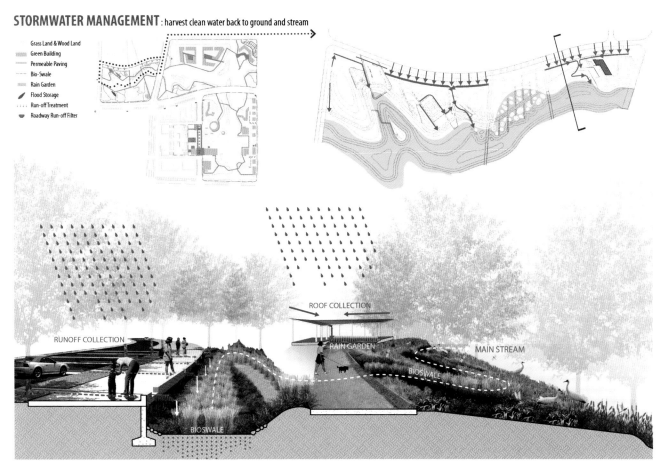

Stormwater run-off is collected and treated before entering the major waterway. This process is demonstrated and included in the park program and design for educational purposes.

UTILIZING EXCESS FILL

RECYCLE SITE CONCRETE TO PERMEABLE PAVING

2006

2009

2012

2014

UTILIZING CONSTRUCTION WASTE SOIL FOR LANDFORMS

2009

2012

2013

154

沃勒溪：城市绿洲

WALLER CREEK: CITY OASIS

项目类型：城市绿道

项目位置：美国德克萨斯州，奥斯汀市，沃勒溪

项目规模：2.4 km

设计公司：土人景观 + Lake | FLATO 事务所（美国）

Project Type: Urban Riparian Corridor

Location: Waller Creek, Austin, Texas, USA

Size: 2.4 km

Design Firm: Turenscape + Lake | FLATO (USA)

沃勒溪修复工程创造了健康而弹性的自然系统。在构建生态基础设施过程中，设计完成了两大主题内容：将沃勒溪转变为绿洲，重构溪岸社区与溪流间真实和共生的关系。

设计采用低技术的方法，即"农民的方法"，具体技术集成为"农民的工具箱"：灌溉工程、土方工程、植被生长、遮阴设施和聚会活动。方案改善了四大公园：从最北端的滑铁卢公园，一路向南到音乐港湾、棕榈公园和小鸟夫人公园。

挑战与目标

沃勒溪坐落于奥斯汀城市中心。河道曾经是城市形象的标志，经年累月，现在却沦落为阴暗的角落：河道渠化导致溪流变窄、退化。季节性洪水和山洪限制了溪岸廊道周边的城市发展。同时，溪流遭受着土壤侵蚀、水体污染和外来物种入侵，使这里的生境遭到破坏。溪流变成了人们避而远之的后杂院，城市与自然被割裂。此外，奥斯汀还受到长期干旱和气温升高的困扰。这次修复为沃勒溪带来了机会：使大自然得到修复，使城市得到修补，重建城市与自然的关系，并使溪谷成为带动周边城区发展的中心。

设计理念

方案将沃勒溪的转变视为一个动态过程，由一定的理念引导、并可适应临时出现的变化。设计采用最小化干预的方法以"加速或减缓自然过程"，达到最大限度发挥生态系统服务功能的目的。设计师像农民一样，依照时序，重整土地、净化溪水、利用周边可持续资源，为沃勒溪营造美丽的景观环境。

设计策略

为使该项目采用的方法在未来项目的更多发展中可以共享，设计师为沃勒溪修复工程筛选了15种简单的方法，构成"农民工具箱"。在柏木大道建好之后，场地上将出现很多宽阔空间和再开发的机会。四大公园各具特色，成为社区民众可参与的资源。方案还根据未来的投资预期为沃勒溪制订了分期发展规划。

1. "安全格局"

为应对洪水和可持续、健康城市生活的挑战，方案构建了"安全格局"（也可以称为"生态基础设施"），作为当地景观中最有效的保护机制。"安全格局"竖向来看，是一叠相互依存的图层，可以保护当地的水文、生物过程和文化遗产。水平来看，其是一个树状网络：沃勒溪是树干，延伸至城市肌理的绿地是树枝，而社区中的场地是小树枝和树叶。

2. 净化溪水和灌溉

源自各个来源的雨水被滞留，通过生态墙、梯田种植带系统得到净化。"智能"雨洪技术通过地下管网系统，耦合水文过程和降雨过程，控制雨水的释放速度，以留住雨水。收集到的雨水、沃勒溪隧道中的再循环水、周边建筑释放的灰水、甚至空调的冷凝水都可以通过灌溉系统得到净化，同时滋润"绿洲"。贮水池和水塔可以留住雨洪和中水，以提供稳定的水源，并起到景观地标的作用。

地表径流分析
Ground surface runoff analysis

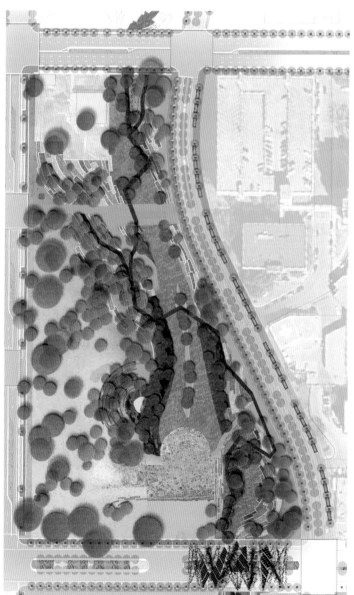

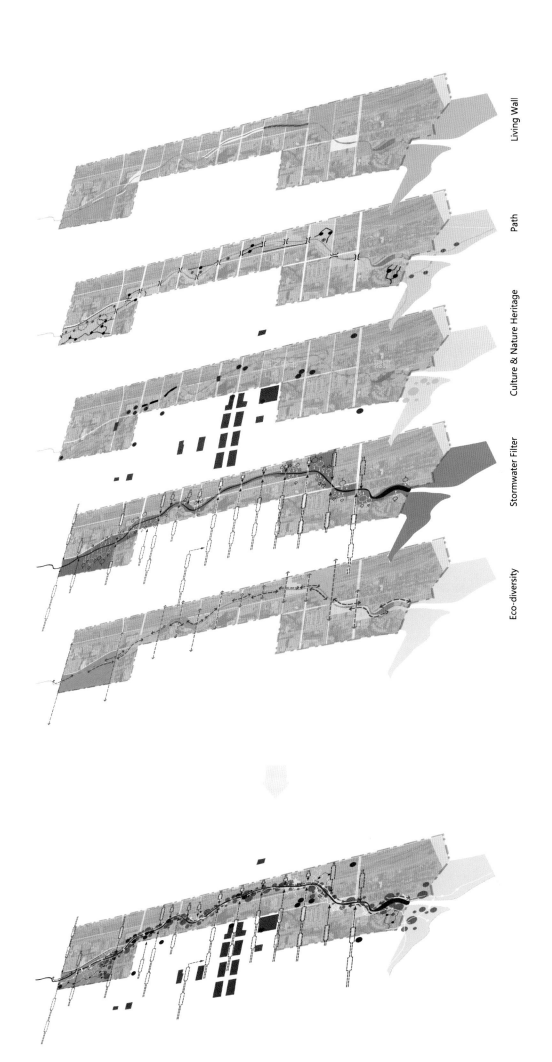

3. 土方工程

通过最小干预的填挖方工程，营造梯田阶地，形成适宜于各种植物生长的微地形和生态墙体。阶地地形将留住坡地土壤、净化雨洪并营造健康的溪岸乡土环境。生态墙的材料来砸掉溪流硬化工程后产生的废渣石材，分为湿润墙体和干燥墙体两种，为过往的行人调节空气。生物护岸工程将减少土壤腐蚀，营造健康的溪岸环境。

4. 植被生长：适应性农田、设计实验和城市农业

受生态修复实践启发，适应性农田能够控制植被入侵、激活修复过程并引发种子被动传播（可以是风力传播、水力传播、动物传播及人类活动传播）。方案中的梯田生境具有生物多样性，是典型的"杂芜的自然"，与人工元素形成鲜明对比。为使将来的修复更成功，设计将沃勒溪视为实验田，以比较不同的种子传播方式的效果（主动传播、被动传播和特定物种杂交）、土质改善和灌溉情况等。城市农业（社区花园、食物、果林）等各种形式相结合，为区域居民生产食物带来体验的快乐。

5. 为居民和生物提供难忘的体验

为应对干旱和高温，绿洲中的大树和遮阳棚为路人提供阴凉的休息之地。设计中简单、灵活的网线遮阴系统，像片片树林，容纳了诸多重要的自然元素。夜幕降临之后，柏木大道亮起的照明让居民在此放心散步。

柏木大道沿溪流蜿蜒展开，修复工程保留下来的大树亭亭如盖。木栈道的材料来自德州沼泽中回收的"水下柏树"。木栈道将被布置在主干步道上，配以良好的遮阴设施。溪岸阶地上下排布了多个入口坡道，保证场地的可达性。设计充分利用每个公园各自地块的显著特征和特有资源，提供适宜的活动类型，它们串连在一起让家庭活动丰富有趣。四大公园的特色如下所述。

滑铁卢公园（Waterloo Park）提供户外林下剧场，还能作为很好的节日用地，并带动周边城市的再开发。

音乐港湾（Music Bend）成为"红河"音乐圈的延伸，设施包括拥有交互型声音环境、乐器租借馆。

棕榈公园（Palm Park）满足社区居民的最大需求——开敞空间，支持家庭活动和滨水活动，包括城市沙滩、排球场、沙滩球场。公园中的浅水水池和灌溉水渠带来凉爽宜人的空气。

小鸟夫人公园（Lady Bird Park）在三角洲景观序列开始之处重新引导视线。这里有园艺和食物烹饪项目，以及观鸟、划船和让孩子动手参与的科技活动。

总结

沃勒溪方案使用"农民工具箱"：灌溉工程、土方工程、植被生长、遮阴设施和聚会场所。通过一系列生态基础设施的构建，设计充分整合廊道功能，让人、生物、自然雨洪管理、城市农业以及其他绿色基础设施功在这一网络中和谐共生，应对奥斯汀自然环境所面临的各种挑战。

总平面图
Site plan

A strong urban open space system is rooted in robust and resilient nature. By growing this ecological infrastructure of Waller Creek Corridor, the Design structures around two intertwined efforts: the transformation of Waller Creek into an Oasis and the cultivation of creekshed communities that express and authentic and symbiotic relationship with the creek.

The designs are built from a simple collection of practices that constitute the "farmer's toolbox": irrigation, earthwork, growth, shade and celebration. Our proposal includes four major park improvements starting at the north with Waterloo Park and working our way south through Music Bend, Palm Park and Lady Bird Park.

Challenges and Objectives

Waller Creek lies within the heart of Austin city. The waterways, once partly defined the plan for the city became negative element over the years: Due to the channelization, the Waller Creek was narrowed and degraded. Periodic inundation and severe flash floods limited development along the creek corridor. At the same time, the creek suffered from erosion, pollution and invasion, making the habitat disconnected from natural seed sources and culture. And the prolonged droughts and rising temperatures are unpleasant climate features afflicting Austin. The riparian gives Waller Creek an opportunity to reconnect with the city physically and culturally and rejuvenate as a centerpiece that revitalizes the development of nearby downtown.

Design Concept

We envision the transformation of Waller Creek as a dynamic process, guided by the concepts laid out in this proposal, but flexible to respond to emergent opportunities. Our approach is to apply minimal interventions that "speed up or slow down the process of nature" to maximize ecosystem services that benefit people and the city. In this way we, like farmers, take on the role of time as we reshape the land, cleanse the water and harness the creekshed's sustainable resources to grow an unforgettable Oasis experience along Waller Creek.

Design Strategies

The design is built from a simple collection of fifteen patterns with the understanding that further development of this project and future projects can draw on a shared practices applicable to Waller Creek, which constitutes the "farmer's toolbox". After The Cypress Walk is implemented, many open spaces and redevelopment possibilities will emerge. Our proposal includes four major parks could come online independently and in any order as resources and community interest take hold. The proposal also describes four redevelopment scenarios that could be catalyzed by investing in Waller Creek.

1. "Security patterns"

To address the flash floods and sustain healthy urban living, "security patterns", or infrastructure systems, is structured as the most effective protection in a given landscape. The "security patterns" can be vertically envisioned as a stack of interdependent layers that protect hydrological and biological processes as well as cultural heritage. Horizontally, the network might be envisioned like a tree with Waller Creek as its trunk, branches of green linking the creek to urban fabric, and the specific and sites of the creek communities as its twigs and leaves.

2. Cleanse the water & Irrigation

Water from different sources will be retained, cleansed through living walls, planted terraces and drip irrigation systems. "Smart" storm water technologies allow for release rates that mimic slower predevelopment conditions, retaining the rain. Reclaimed wastewater, recirculation water from the Waller Creek Tunnel, gray water from surrounding buildings, and even air conditioner condensate will be cleansed while irrigating support the Oasis. The cistern and water tower will store the storm water and reclaimed water to provide steady base flows and act as markers in the landscape.

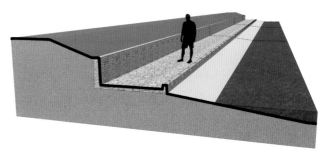

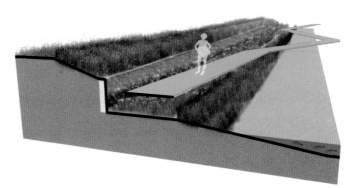

3. Earthwork

Through minimal cuts and fills along the bank, the terraces stretched from subtle microterracing to bioengineered walls. The stepped topography keeps soil on the slope, treats storm water and creates healthy native creekside habitat. Wet and dry living walls will be constructed from the removed hardscape and sited to provide natural air conditioning to passersby. Bioengineering will prevent erosion and shape a healthy creek environment.

4. Growth: adaptation palettes, designed experiments and urban agriculture

Inspired by restoration ecology practices, the adaptive palettes will control invasive, activate restoration and allow passive propagation. They will drive an emergent palette that is diverse and symbolize "messy nature", which will then be contrasted with artful elements. In order to further research into how restoration practices can be more successful, the propose envisioned the Waller Creek a backyard laboratory to compare a variety of active and passive approaches to specific species mixes, soil improvement, irrigation, and more. Urban agriculture forms such as community gardens, food forests and other innovative. Collaborative forms of urban agriculture can provide products and pleasures for area residents.

5. Grow an unforgettable oasis for people and wildlife

In response to the droughts and rising temperature, the oasis provides big trees and solar canopies as effective shades. The proposal's simple and flexible geometries were developed to include most important existing natural assets like mature trees. After dark, the Cypress Walk will remain illuminated for families to stroll.

The Cypress Walk will feature a meandering walkway along the creek, under the shade of the restored riparian forest. Wooden platforms will be built from "sinker cypress" reclaimed from Texas bayous. These platforms will belocated in comfortable shaded locations along the main walkway. Access ramps will wind up and down the terraced banks to ensure accessibility. Each park will have a unique play focus so that together they provide for wide range of family-friendly activities and take advantage of the distinct cues and resources of each place.

The four major parks improved afford different features.

Waterloo Park provides an outdoor theater under the trees, and will be even better for festivals and stimulate redevelopment.

Music Bend becomes an extension of the Red River music scene. It will have interactive soundscapes and an instrument lending library.

Palm Park offers public most critically needed open space for families and water playscape, with urban beach for volleyball and sand play. The wading pools and acequia cools the air off.

Lady Bird Park will reorient attention on the delta where the landscape opens up, and will have gardening and food preparation programming aswell as birding, boating and hands-on science opportunities for kids.

Conclusion

The Waller Creek proposal engaged in the "farmer's toolbox": irrigation, earthwork, growth, shade and celebration. By growing this ecological infrastructure, the design integrate the corridor for people and wildlife, natural storm water management, urban agriculture, and other green infrastructure functions into this network to see Austin through the environment challenges that lies ahead.

164
上海嘉定新城中央绿色纽带："紫气东来"可持续设计

A GREEN TIE AT THE CENTER OF THE NEW CITY: SUSTAINABLE DESIGN OF SHANGHAI JIADING NEW CITY LANDSCAPE AXIS

项目位置：中国上海市
设计公司：Sasaki事务所

Location: Shanghai, China
Design Firm: Sasaki Associates, Inc.

设计理念

"紫气东来"的设计理念围绕可持续发展的景观规划目标。作为一个夹在新城中央的条形绿地,它肩负了大型城市公园的多种使命——既需要服务于周边的人群,又是整个城市生态系统中的重要节点,对整个城市的可持续性有着深远的影响。设计将多元和动态的空间体验与基地的自然环境和嘉定丰富的文化传统相结合,犹如"林中的舞蹈",在嘉定新城的中心打造一条独具个性的绿色纽带。

"林"的概念表现在它不仅成为一个游人众多的活跃场所,还肩负着恢复当地的自然系统和重建野生动物栖息地的重任。场地中栽植了大面积的乡土植被,包括树林、湿地、草甸等。嘉定位于长江三角洲冲积平原,湿地是该地区极具代表性的自然景观。在长期的农业实践中,湿地被转化为农业用地。人们开凿运河排除湿地中的水分,便于将土地用于耕种。运河系统逐渐扩张,成为当地主要的灌溉和运输途径。用地的转化造成过量的水土流失,水质恶化。设计充分利用将这些农业用地转化为大型公共绿地的机会,大量恢复原有的湿地和伴生林地,以修复本土的野生动物栖息地,从而改善生态环境,并营造更加健康的人居环境。

"舞"的出现受到中国传统绘画、书法和舞蹈的启发。自然中的行云流水启迪了传统舞蹈和书法艺术的形式,反过来,这些传统艺术形式又在现代景观设计中得到回应。嘉定当地著名画家陆俨少的绘画极大地启迪了"紫气东来"的设计语汇。设计中的四条走廊交织变化如行云流水,又如水袖飞舞,沿空间和地形变化蜿蜒起伏,既承载着重要的交通功能,又将园内的各项用途整合成一个整体。

在设计理念的统筹下,景观中的空间构成涵盖了开敞与封闭、大型与私密、活跃与安静、城市与田园、直与曲、高架与下沉等多种变化。所有空间由四条蜿蜒的走廊串联起来,通向周边的开发地块和公园之外的其他绿地。地形和植被的起伏变化,不仅使游客从不同的高度欣赏周围的景色,得到丰富的视觉体验,也为野生动物和行人提供了过街走廊。"林中的舞蹈"整合了公众、社交、商业、文化休闲和生态恢复功能来,以营造城市中心绿地。

园内的活动内容与周边毗邻的用地功能相呼应,分成五个区:社区活动区、健身区、政府和科教中心区、交口茶座区和湖区。每个区域都有自己的独特之处,为毗邻的区域以及整个嘉定新城服务。公园里的活动内容多种多样,以吸引不同年龄、不同兴趣和不同日程的人群,使公园成为一年四季、一周7天、一天24小时都对当地居民和外来游客极具吸引力的地方。

雨水管理系统

雨水管理系统包括雨水的净化、收集和重新利用。项目中的大面积湿地、沿路的生物过滤区和沿河的植物过滤带具有净化和吸收地表径流及净化运河水质的作用,公园里的湿地面积足够处理来自公园全区、周边和内部道路以及东云街开发地块的雨水径流。水花园处的雨水收集箱收集和过滤道路上的雨水,将其引入水生植物池进行处理后再引入泡泡池,作为水花园的主要水源;在林荫走廊附近的两个地下储水箱可以收集林荫走廊上的地表径流,经过过滤,用于浇灌公园中仅有的两块草坪。

基地的水质问题来自运河上游的污染、新开发地块和道路上的地表径流。运河在到达基地之前穿过大面积农田、村庄和工业区。由于农业中大量使用化肥和农药、农业生活污水未入管和工业污水排放缺乏管理等,大量的营养物质和工业废物排入运河。新城中传统的雨水管理系统集中收集道路和开发地块上的地表径流,未经处理直接排入运河,将大量重金属和多环芳烃带入运河,加剧了运河水质问题。原有的运河两岸又多为硬质直立驳岸,不但对水质改善没有提供任何帮助,而且影响了两栖动物和涉禽建立栖息地。在公园建造之前,运河水属于IV水质。

在公园建成后,河道水质得到了极大改善,沿河湿地也成为鸟类和其他野生动物的天堂。

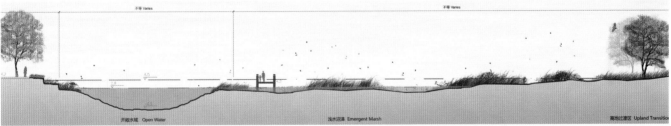

Design Concept

The design concept of Ziqidonglai is created around sustainability. As a linear green space placed between developments at the center of the New City, it carries multiple functions of a large urban park-it serves the communities around as well as functions as key node of the urban ecological system-which creates tremendous influence on the sustainability of the whole city. The design integrated diverse and dynamic spatial experience with the natural settings of the site as well as the rich cultural background of Jiading, to create a unique green tie at the center of Jiading New City, like "Dancing in the Woods".

The "Woods" part represents this park takes mission to restore the natural system and reestablish wildlife habitats on site, while creating an active place for people. Native plant species are planted in large area of woodland, wetland, meadow, etc. Jiading is located in the flood plain of Yangtze Delta, where wetland is its original landscape. In the long history of agriculture, wetland was gradually turned into farm land. People dug canals to drain wetland and create more usable land for agriculture, and then canal system was expanded for irrigation and transportation. The land conversion creates excess runoff and deteriorated water quality. We want to take this opportunity of converting farm lands into large public green space, to recover the original wetland and associated woodland, to recover the local wildlife habitat, therefore improve the ecological condition and create a healthier living environment.

"Dancing"concept was an interpretation of traditional Chinese painting, calligraphy and dance. Natural landscape elements such as floating cloud and flowing water have inspired traditional dance and calligraphy. Conversely, the flowing and dynamic forms of these traditional arts translate well into the contemporary landscape design. The works of Yanshao Lu, a well-known local artist, has greatly inspired the design vocabulary of Ziqidonglai. Four major paths in the park interweave and interact with a variety of park elements, as choreography of movement, twisting and turning along the space and landforms, makes the whole composition. They are part of the circulation system as well as links with various uses in the park.

With this design concept, the spatial configuration of the park embraces the variation of open and close, celebrational and intimate, active and serene, urban and pastoral, straight and curvilinear, elevated and sunken. All the spaces are linked by meandering paths connecting with surrounding neighborhoods and major landscape nodes beyond the park. Undulating landform and vegetation offer rich visual experiences at different elevations while functioning as wildlife and pedestrian corridor crossing major roadways. "Dancing in the Woods" integrates public, social, commercial, cultural, recreational uses and ecological recovery to create the central green space of the city.

The programs of the park respond to surrounding land uses by divided to five zones: Community Zone, Fitness Zone, Government and Interpretive Center Zone, Crossroads Café Zone and Lake Zone. Each zone has its

distinctive character, and serves adjacent development and the entire Jiading New City. Diverse park programs are to attract user groups of various ages, interests and schedules, and make the park a destination for local residents and visitors a year, a week and a day round.

Stormwater Management System

The stormwater management system was designed to include rainwater mitigation, capture, and reuse. Large wetland, roadside biological filter area and riverside planting zone in the project help clean up and absorb runoff therefore improve the canal water quality. The area of wetland is adequate to treat stormwater runoff from all park area, frontage and internal roads, as well as buildings in Dongyunjie development. The rainwater collection and treatment structure in the water garden area capture and filter rainwater from the roads. Filtered water is led to aquatic plant pools for further treatment, and then to bubbling pools as the main water resource for the fountain. The two underground water storage tanks adjacent to the Tree Path collect and filter surface runoff, which is then used to irrigate the only two patches of lawn in the park.

The water quality issues on site involve both existing canal water bodies and the stormwater runoff from new developments and roads. The canal water runs through extensive agricultural field, villages and industrial area from the upstream. Due to the extensive uses of fertilizer, pesticide, lack of sanitary sewer system and industrial waster water management, the canals carried large quantity of nutrients and industrial wastes. The traditional stormwater management system at the new city collects the runoff from streets and developments and discharge it directly into the canals, which brought heavy metals and PAHs into the canals and worsen the water quality. The existing vertical embankment was not helpful in mitigating the water quality, neither providing habitat opportunities for amphibians or wading birds. The water quality at the canals was at Level IV before the park was built.

After the park was built, the canal water quality was greatly improved, and the wetland alongside has become a heaven for birds and other wildlife.

170

吴淞滨江净水公园

WUSONG RIVERFRONT WATER TREATMENT PARK

项目位置：中国上海市
项目规模：95 ha
设计公司：SWA集团
翻　　译：张旭

Location: Shanghai, China
Size: 95 ha
Design Firm: SWA Group
Translated by Zhang Xu

设计团队所构想的提案是设计一套可解决即时水体污染问题的净水景观基础设施。第一阶段是将水质作为基本关注点，以期冀水质改良起到反哺作用，包括重新唤活植被生机、创造沿岸栖息空间、融合娱乐和社交项目。该项目的终极目标是将先前退化弃置的水网系统打造成宜人的景观基础设施。自被授予此项任务之日起，领队景观设计师与协同咨询人员（包括水质科学家、湿地生物学家、水文工程师、建筑师和开发商）对场地进行了为期一年的综合分析。占地面积95 ha的第一期基地位于极其关键的牛轭湖地段，随后开始建设上游的净水试验区域。

将设计与规划过程相结合：在基址调研中发现场地内有很多先前砖厂遗留下的挖掘坑基，周边区域地块上形成的未经处理的表面径流将这些坑基淹没，并且针对该滨水项目的大部分内容而言，并不适宜直接取源于河水。在项目中，将水流速度设计为可将水质由Ⅴ级净化致Ⅲ级的程度。考虑到湿地自身净水能力的限制，设计运用静态水处理技术（湿地净化处理）与动态水处理技术（保证曝氧池中的空气质高量足），实现净化水质的目标。

作为专项分析与设计流程中的有机组成，现处建设阶段的公园也隶属于项目试点系列。专门检测建筑不同部位的信息收集汇总之后，被用于指导与优化未来的运营阶段。这种方法是该项目的关键之处，对于水的输出补偿均衡、水压控制、植被养护、实践季节性变化景观都举足轻重。在规划设计中结合设计建造思想能够可最终的建设项目质量，丰硕最终的建设效果。

该工程位于未来发展地块的上游，并用河水与市政排水作为补给。该系统模拟了一系列自然过程，功效形同河流的"肾脏"，可以移除河道上游的淤泥与工业废水，使河道下游甚至更广范围均能受益。一系列水池、河道在有氧厌氧交替的环境中通过沉淀、过滤、曝气、生物作用将靶转向污染物移除。

为了实现最终目标，该目标导向型的水力学设计工程进一步在技术层面对景观营造做了一番探索，包括对滞留时间和水流速度的研究。在进行评估分析后控制水流的流速与流量，以此避免河道的淤堵。设计也提出了"式样翻新"的举措，根据基地现有的高程状态布置净水结构单元，出于保持土壤的目的平衡填方与挖方。例如，保留挖掘坑池用于曝气过程，水流通过最长的路径流经区分出等级的净水水道，通过湿地植物的入渗作用完成过滤。除了净化过程是经过精确计算之外，净水体系也具有兼顾适应洪涝与干旱状态的弹性。植被设计是依据植物的空间特性与短期序列策略，因为净水功能较强的植被往往需要富养性水质；当水质得到改良后，其需让位于其他植被品种。

新型净水处理系统的设计同样也注重用户体验并强调公共教育功能。净水湿地公园中，在最初的试点基础上，出于对功能效用的考量，将池塘与水网攒簇以构建花园与开放空间。例如，沉降池兼具景观映景池的作用，净水水道装点了石景园并用作鸟类栖息通勤廊道，除此之外，曝气过程在经过艺术性处理之后，通过涟漪层层、泡沫丰富的景观池展现出来。漫步走廊贯穿整个净水湿地公园，把功能迥异的项目空间与各具特色的景观风貌串联起来，将它们各自的故事编织入水质净化的奇妙旅程。就这个层面来说，该设计也不乏是一项寓教于乐的启智体验。不再使用封闭管道式的工程处理方式，使社区居民可调动自身的视觉感官亲身体验水质净化过程。

引入水质净化体系，对于滨河空间的开发而言有重要作用。将湿地建造技术引入已建成区域，扩展了人们当下对景观设计的认识。经过设计的景观由静止的装饰脱胎换骨成可提供生态服务、促进场地优化的复杂系统。随着该试点项目的竣工与落成，周边地块的地价也得到了大幅度的增长。

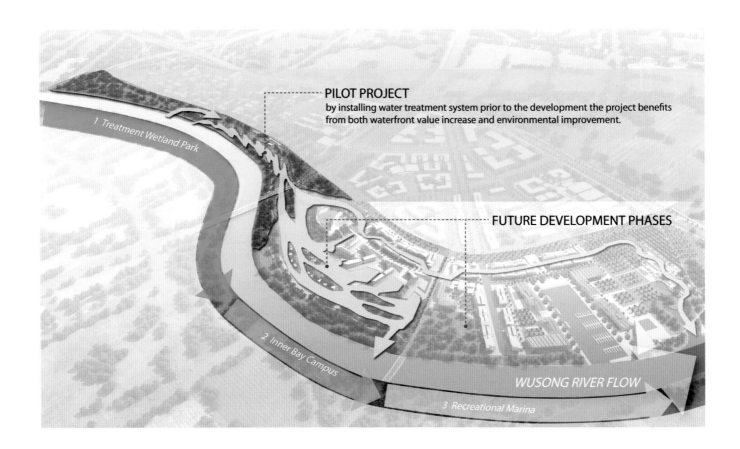

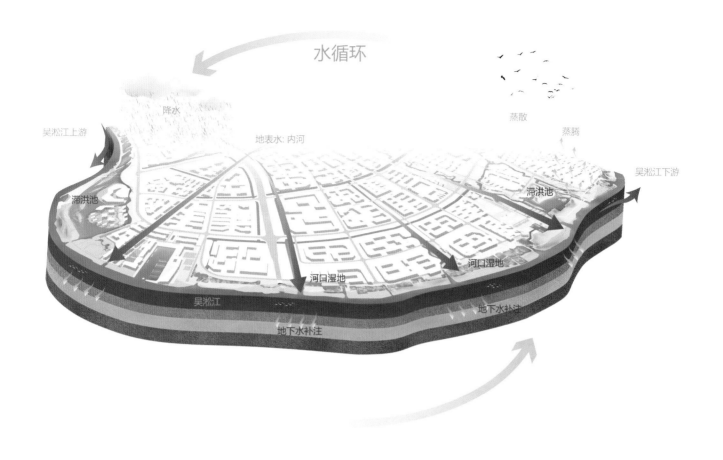

水循环

降水 | 蒸散 | 蒸腾
吴淞江上游 | 地表水：内河 | 吴淞江下游
滞洪池 | 滞洪池
河口湿地 | 河口湿地
吴淞江 | 地下水补注
地下水补注

水质改善
80%以上
80% IMPROVED

BOD | N | NH₃ | P

2008 — 2013

昆山花桥服务外包区，致力于提供区域产业服务配套、休闲设施，并研究科普改善水质，根据2013年资料，水质净化由2008年的五类水提升到二类水，甚至一类水的标准。

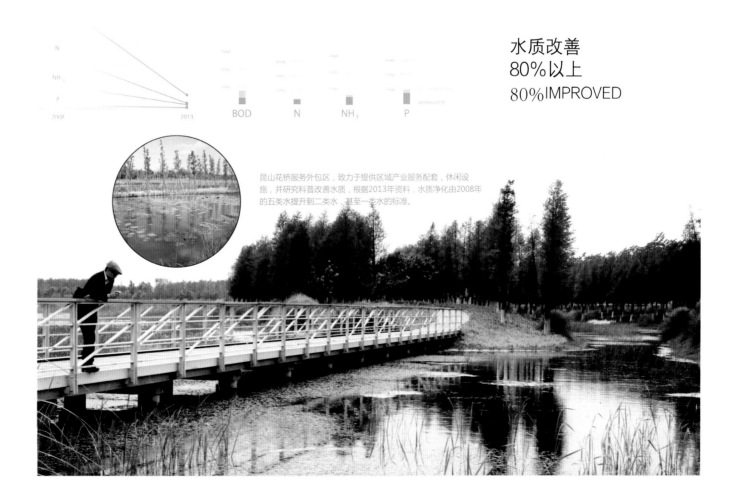

The design team's proposal envisioned a water treatment landscape infrastructure to address the immediate issue of water pollution. The first phase made water quality the primary focus, suggesting that improved water quality would in turn make possible re-vegetation, habitat creation along the banks, and integration of recreational and social programs. The ultimate goal was human habitation of a formerly degraded water network. After being awarded the commission, a year-long comprehensive analysis was undertaken by a collaboration between the lead landscape architect and affiliated consultants, including water-quality scientists, wetland biologists, hydrology engineers, architects and developers. The 95-hectare first phase at the key oxbow portion then started its construction at the upstream treatment pilot area.

Integrated Design and Planning Process: The site study found that the site was riddled with excavation pits left by a former brick factory, inundated by untreated surface runoff from adjacent parcels, and the river water was unsuitable for most waterfront programs. The design established a treatment flow rate capable of treating Class V water to Class III water. Recognizing the limits of wetlands alone for treatment, the design integrates passive (treatment wetland processing) and active (fine diffused air in aeration ponds) water treatment technologies to achieve the water quality goals.

As part of the unique analysis and design process, the park is currently being constructed as a series of pilot projects. The importance of the pilot project approach is that information gathered from site specific monitoring of built portions informs the refinement of future stages. This is especially critical for determining water budgets, pump controls, plant establishaent and anticipating seasonal changes. Linking the planning process with design and construction thinking strengthens and enriches the eventual built project.

The pilot project is sited upstream of future development parcels, and intakes both river water and municipal stormwater discharges. The system mimics a wide variety of natural processes and acts as the "kidney" for the river, removing sludge and industrial effluents discharged into the river upstream, extending the benefits of the park downstream to a larger region. A sequence of pools and channels remove targeted pollutants through settling, filtration, aeration, and bio-processing in alternating oxic and anoxic environments.

This goal-oriented hydraulic design/engineering further explores landscape based techniques for achieving treatment objectives, including estimating residence time and flow rates, manipulating velocity and volume through grading, and avoiding stagnation. A 'retrofitting' approach to design was proposed, in which the water treatment cells were designed to fit into the existing site contours, balancing cut and fill for soil conservation. For example, excavation pits were preserved and utilized for the aeration process, while treatment channels were graded to let water filter through wetland plants over the longest paths. While the purification process was precisely calculated, the system is designed to be flexible to accommodate flood and drought conditions. Planting design derived from the plants' spatial qualities and a temporal succession strategy whereby cleansing plants, requiring high-nutrient water, would give way to other species once water quality had improved.

The design of the new water-treatment system also considers user experience and emphasizes public education. In the Treatment Wetland Park, built upon the initial pilot project sites, ponds and channels are conceived as a series of gardens and open spaces, based on functional uses. For example, a sediment pond is also a reflection pool; a treatment channel becomes a stone garden and bird blind lounge; and the aeration process is artistically expressed as ripple and bubble pools. A promenade runs the

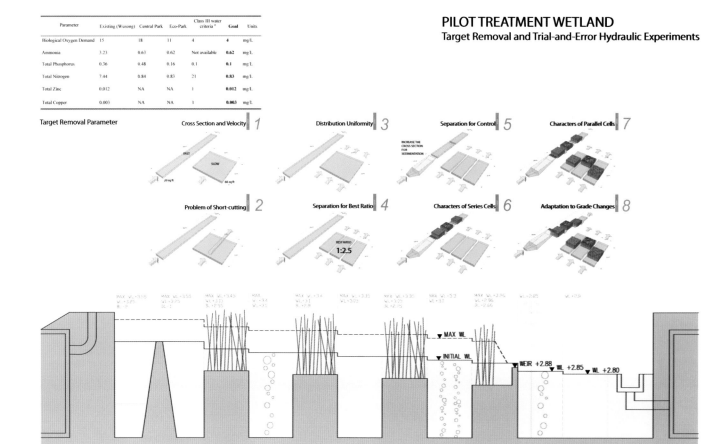

PILOT TREATMENT WETLAND
Target Removal and Trial-and-Error Hydraulic Experiments

length of the Treatment Wetland Park, connecting a variety of programmed spaces and distinct landscapes, and weaving together the story of the water-purification journey. In this way the design is an educational experience. The community can witness the process of water cleansing, in a much more visible manner no available with typical closed pipe engineering solutions.

The water cleansing system will serve as a model for responsible development along the river, introducing constructed wetland technology to the region in a built form and expanding current perceptions of designed landscapes from passive ornament to active, complex systems capable of providing ecosystem services and enacting change. After the installation of pilot project, the surrounding parcels experienced significant gain in property value.

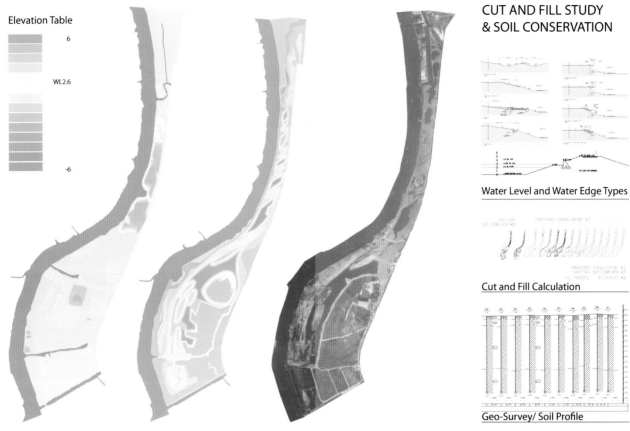

CUT AND FILL STUDY & SOIL CONSERVATION

Water Level and Water Edge Types

Cut and Fill Calculation

Geo-Survey/ Soil Profile

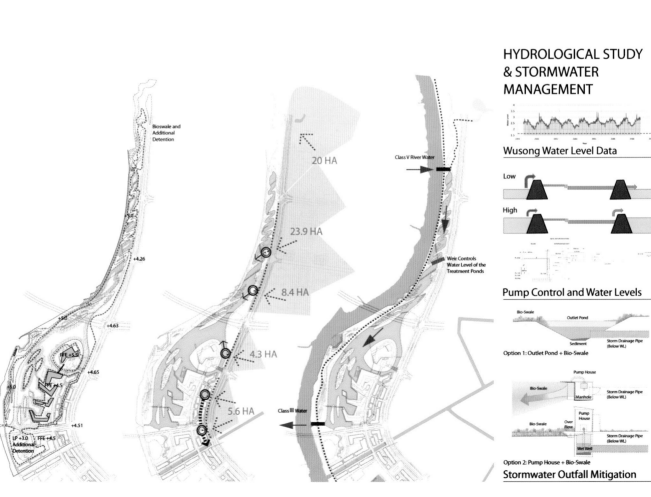

HYDROLOGICAL STUDY & STORMWATER MANAGEMENT

Wusong Water Level Data

Pump Control and Water Levels

Stormwater Outfall Mitigation

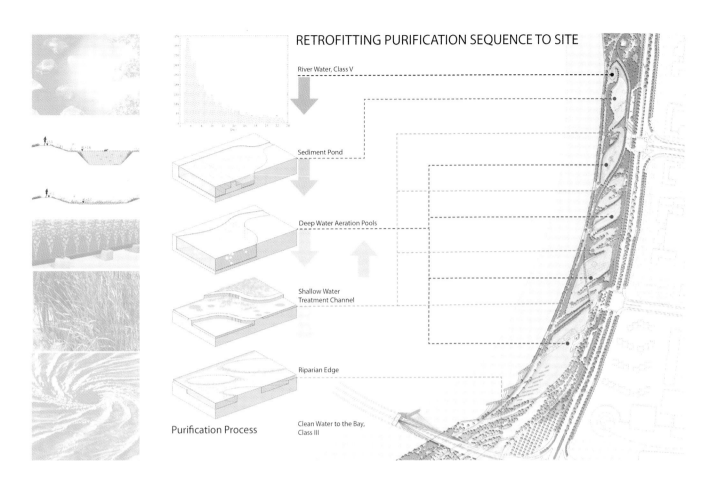

RETROFITTING PURIFICATION SEQUENCE TO SITE

- River Water, Class V
- Sediment Pond
- Deep Water Aeration Pools
- Shallow Water Treatment Channel
- Riparian Edge
- Clean Water to the Bay, Class III

Purification Process

178

深圳市水土保持科技示范园
SHENZHEN SOIL AND WATER CONSERVATION PARK

项目位置：中国广东省深圳市
项目规模：50 ha
设计公司：深圳市北林苑景观及建筑规划设计院有限公司

Location: Shenzhen, Guangdong Province, China
Size: 50 ha
Design Firm: Shenzhen BLY Landscape & Architecture Planning & Design Institute Ltd.

蚓之丘：模仿蚯蚓在土壤中拱起的洞穴，通过放大、形象的人工塑造，让观众在洞内感受蚯蚓在土中穿行的过程，揭示自然界动物疏松、改良土壤的活动。

深圳市水土保持科技示范园选址于深圳市南山区乌石岗废弃采石场，占地面积50 ha。在此废弃采石场中，一方面因地制宜、因山就势利用乡土原生与废弃材料，进行场所的景观与生态修复改造，另一方面充分研习自然的山水路径，模拟自然表皮对水的吐纳呼吸，打造集展示、教育与实验、科研于一体的户外课堂和开放式公园，实现废弃场地的复兴和激活。

深圳地处低纬度地区，紧临南海，暴雨频繁，洪涝灾害易导致巨大的损失，项目通过全面完善园区低影响开发雨水系统，将透水路面、生态草沟、雨水花园建设等理念贯穿水保园建设，并突出水土保持科研与展示，使水保园成为全国首个突出"海绵城市"城市水土保持特色的科技示范园区。

"城因水丰土沃而兴，城随水枯土瘠而灭"。深圳作为中国改革开放的窗口和最先进行城市水土保持实践的城市，曾经大规模的城市开发建设和不合理的推山造地，水土资源一度遭到严重破坏，水土流失加剧，也极大地干扰了水源保护地的地表径流，增大了地表流的流速，破坏了原有汇水区。深圳市水土保持科技示范园所选址的乌石岗废弃采石场正是遗留下来的典型水土创伤之一，也是原有自然水土海绵的破损处之一。在这样一块场地上建造水土保持科技示范园，不仅实现了景观生态修复，弘扬了水土文化，展示了水土保持技术，也给未来深圳的发展提供了海绵城市的发展方向、低冲击开发的技术路线。

根据场地现状汇水分区情况，统一布设一、二期雨水体系，以低影响开发模式，把场区分为"坡""谷""路""园""建筑"区块，分别处理，通过源头分散控制、径流蓄排结合、末端集中处治等措施，延长雨水入渗时间，减少水土冲刷强度，提高场地水土涵养，建立水土保持综合生态治理模式，实现场区内雨水"零外排"。

恢复采石场时期被破坏的汇水区，形成水清园、黑土沼泽等雨水洼地，沿溪谷、山脊线设置汇水湿地、草沟等，划分不同的汇水面，通过种植设计保证良好的下渗和滞留效果；形成雨水在场地中的蓄、滞、渗、净、用、排全功能路径，构筑完整的点—线—面体系完善的海绵花园。

在优美的山林自然环境中有序地设置了各种城市水土保持科普展示设施。区别于教科书的刻板教学，通过装置改进、亲手操作，将户外教育、互动参与、模型展示紧密结合，大众可清晰直观地了解水土资源对于人居环境的重要意义。

项目规划在园内设置了中国第一家融合城市水土保持和海绵城市理念的科普教育4D影院，针对中小学生和普通市民，通俗地讲解发生在身边的故事，再通过声、光、电等手段的再现，使观影者感同身受，从而激发其对水土资源保护的思考，进而亲身践行各种绿色低碳行动，实现理想城市的最终诉求。

项目规划全过程以"景观水保学""海绵城市"等新兴学术理论为指导，通过景观、生态、水保的综合手段多维度实现场地的复兴。这里不仅是一个环境优美的公园，更是一个体会人与自然和谐之道的平台，其可增强市民"敬畏自然、尊重土地"的意识，最终实现"土返其宅，水归其壑，草木归其泽"的生态文明之梦。

总平面图
site plan

Shenzhen Soil and water Conservation Park is situated in Wushigang disused quarry in Nanshan District, Shenzhen. It covers 50 hectare. In this disused quarry, abandoned material, buildings and other constructions are used in restoration of the place. Part of the quarry's original appearance are kept to let visitors experience personally, having an effect of caution on them. At last, it becomes an outside classroom and free park, integrated with displays, education, experiment, scientific research and landscape scene.

Shenzhen is located in low latitude region, next to South Sea. Heavy rain frequently happened, and cause huge damage to the city districts. The project generally improves the low impact rain water system, by permeable pavement, bio-swale, rain garden and other theories. It mainly presents soil and water conservation technology, making the park the first soil and water conservation park with "Sponge City" theory.

Cities flourish because of abundant soil and water, and die because of impoverished soil and water. Shenzhen as a window of the reform and opening-up carries out soil and water conservation at first. At one time, because of its large-scale urban development and construction as well as irrational pushing hills into lands, Shenzhen's soil and water resources had been once ravaged with worsening water loss and soil erosion. It also interrupted the surface runoff, increase and flow rate and destroyed the original catchaent area. Shenzhen Soil and water Conservation Park realizes landscape and ecology revitalization, promote soil and water culture, presents technology of soil and water conservation, and also provides direction for sponge city construction and low-impact development.

There is a overlook arrangement of first and second phase of rainwater system according to the catchaent area. The site is separated in slope, ravine, road, garden, architecture and implemented in low impact development. By dispersed control of source, flow line storage and treated in the end side, it can prolong the time of rainfall penetration, weaken water and soil erosion, build a overall soil and water ecological management, to realize no rainwater efflux in the site.

The original water catchaent area is recovered and forms Water Garden and Black marsh rain depression. Along the ravine and ridge line, there are different level of catchaent wetland and grassed waterway, to ensure good infiltration and retention by vegetation. The site has constructed a fully functional path of rainwater impounding, retention, infiltration, purification, usage and efflux, to build a sponge park with perfect system from dot to line and surface.

Science and Technique District of Soil and water Conservation set place in elegant forest and install varied display of soil and water conservation in different cities. Different to school teaching, it emphasizes improvement of equipment, personal operation, and combines outdoor education, interactive participation and model display together. It makes the public understand the importance of soil and water to living environment clearly and directly.

Jin Zhe Garden, reconstructed from deserted buildings, is the first 4D Technological Cinema based on soil and water conservation and sponge city idea in China, specially used for playing science and educational film about water conservation. Aiming at primary and middle school students and ordinary citizens, with easy-to-understand explanations, stories take place around and technologies of sound, light and electricity to reappear them make visitors count as their personal favor. Then, they will be aroused to concern about soil and water conservation, take green and low-carbon practice, and realize the final appeal to ideal city.

Planning and design based on the new theories of Landscape Soil and Water Conservation and Sponge City, realize the site revitalization through comprehensive means of landscape architecture, ecology and soil and water conservation. This place is not only a gorgeous park, but also a platform to comprehensive harmony of human and nature. It hopes to motivate and increase people's idea of respect nature and earth, and to finally realize the ecological dream of "earth back to its house, water to its ravine, and vegetation to its river bank".

总平面图

- ① 生物滞留池
- ② 生态草沟
- ③ 前置塘
- ④ 雨水花园
- ⑤ 观景平台

设计说明

本区域是海绵城市【蓄】科普示范点之一；
① "生物滞留池——对雨水进行截流过滤，其形态为下沉绿地，平时兼做紧急疏散空间；
② "前置塘——沉泥区，雨水经过该区域时泥沙沉淀，然后进入雨水湿地。
③ "雨水花园——超过调蓄水位时，雨水经过溢流管排入水库。

海绵城市生物滞留池
Sponge city bio-retention pool

沟道治理展示区及生态排水沟
Channel & tunnel management display area and ecological drainage trench

以"五色土"的概念展示不同地域的土壤的持水率、含水率和在雨水环境下的变化
The concept of "five colored earth" demonstrating different regions with various water holding ratio, water contain ratio and the variations of rainwater in wet environment

锈钢板与白沙砾径和原有植物相映成趣
Stainless steel, white gravel paths and original plants full of reflections with each other

结合边坡陆槽设施,展示雨水在不同下垫面下的流速和下渗率
Combined with steep-edged slope & groove facilities, displaying the various rainwater flow rate and infiltration rate in different underlying surface

废弃采石场遗址展区
Display area of abandoned quarry

生态石砾路
Ecological stone-grind road

生态石砾路
Ecological stone-grind road

186

长沙巴溪洲景观规划

LANDSCAPE PLANNING OF CHANGSHA BAXI ISLAND

项目位置：中国湖南省长沙市
项目规模：73 ha
设计公司：SWA 集团
翻　　译：张旭

Location: Changsha, Hunan Province, China
Size: 73 ha
Design Firm: SWA Group
Translated by Zhang Xu

项目目标

项目规划力求模拟河流生态的力量与弹性，打造一个大尺度的公园，营造变化万千的生态环境。

巴溪洲是湘江三岛屿之一，临近长沙市中心。"巴溪洲"的内涵是"沙洲"，用来形容这个岛屿颇为贴切。江水的力量瞬息万变，塑造了巴溪洲。较宽的河道与较快的流速导致河岸东沿被腐蚀，而西岸自然湿地的形成则归功于窄小的河道与较低的流速。在自然造化中，河流岛屿滩涂的形状与河流的流向和河水的运动状态息息相关。季节性的洪泛形成了河流的内道，岛屿上的植被也可反映出土壤状况与水位高低。

根据设计构思，为了适应季节性水位线的波动，巴溪洲被塑形成一系列的线性土丘与梯田。其具有方向引导性的地形使水位升高的河水流经或流过岛屿。岛屿的周边地带成为植被物种丰盛的湿地，并用作保护土壤免受河水冲刷的缓冲区域。绵延连展的湿地梯田展现了当地风光旖旎的稻谷农产景观。当水位上升时，山丘上的森林绿植会代替湿地植被，其中绝大多数都是本土的树木与花草。营造地势更高的梯田是为了给公共设施、亭子和博物馆提供更好的落址地点。南部湿地区域中，湿地岛屿和小型湿地半岛相互依傍，蜿蜒的漫步栈道体系穿插其中。它们的高程较岛屿稍低，在丰水期可因洪泛作用形成长草沼泽，并为各种湿地植物的生长繁茂提供支撑。

湿地梯田让岛屿有血有肉、形象生动，并且在很大程度上使它免遭水蚀。小径和步道贯穿其中。漫步其中，游客到达亭台、瞭望码头、露天广场、停泊区、湿地影院抑或雕塑花园，并在乡土水杉林下邂逅别墅小院。

发现之旅

人工栽植的本土水杉林成为长沙岛的绿色脊梁，一系列花园和绿地也因此在周边蔓延开来。由小径、步道和自行车道组成的广袤系统为各种各样的探险活动提供了便利。"龙花园"追溯了长沙悠久的历史，一股清泉在巨石间喷涌而出，一方石质马赛克拼画矗立其间。另一个由纤维和铁丝绑扎而成的"飞龙"在树木之间穿梭游走。一处静谧的茶室隐迹于巨大的竹园之中，因此濒临一池菌苔，承袭一片林缘飞地。一个巨大的树屋坐落在儿童乐园中，秋千滑梯一应俱全。池塘片片，是小型水生植物的补充栖息空间。在展览馆中，游客可听到关于这座古老水城怎样发迹成为华夏中部交易通埠的故事。曾几何时，这里也是"中原王朝"的畿内首府。游客穿过延绵数里的小径可漫步至静谧宜人的草地与树影婆娑的森林，这里是长沙城市的留白，游客可在这里冥想休闲、返璞归真。在这里，自然造化协同科学技术，在城市的中心缔造了一方崭新的滨河绿洲。

保持岸线稳定

岛屿东岸边缘水流湍急，岸线侵蚀已属常态。为了稳固岸线，该项目在此建造了一批半下沉式工程筑堤，以保证湿地梯田在滩涂线内顺利搭建。持续移除浅根性水生草丛，取而代之的是芦苇、香蒲等深根性湿地植被，它们植被是东岸腐蚀力度最强地段的绿色屏障。在高水位时期，该线状长草网络将会降低水流速度，使东岸边沿免受侵蚀。同样，草木葱茏的河岸也可为鸟类与野生动物提供良好的栖息环境。

西岸表面受低缓水速影响，沉淀作用显著。因此，自然湿地是线状中的主要景观。为了保护并拓展这些湿地，不但在此基础上营造了蜿蜒的湿地梯田，而且在不同的高程之上改造斜面坡度以应对季节性的洪泛。

夏季洪水期被淹没的岸线
Shoreline –submerged by summer flood

从历史角度来讲,像湘江这种河流生态系统,河岸周边应该会有大量木头堆积。这些木头是被大型洪水冲倒的树木,随着水流对岛屿与河岸物质的冲刷和搬运作用漂移至此。河流系统中的残存木头发挥了稳固岸线的关键作用并且为野生动物提供了重要的栖息场所。为了在生态恢复设计中对这些木质碎片善加利用,南部湿地无论从结构层面还是在栖息层面都为木片的再利用提供了至关重要的契机。为了引导水流进入河道远离岸线从而减少冲刷,部分工程木质阀门被置放在南部(位于东岸)草皮护坡湿地水道的起始部位。这些结构在洪涝来临时可削减洪水的冲力,并保持岛屿结构的完整性以及湿地浅草与灌木植被群落的物种多样性。

水位波动

河流每年的水位波动范围是30 m到36 m。经过设计的土丘与梯田可使河水流经或流过岛屿以应对季节性变化。景观设计在不同的水位线位置种植不同的植被群落以适应河流水位高低的波动。

根据现有的岛屿地形地貌划分出不同的设计等级,在整个岸线长度范围内将挖方作业与填方作业进行最小化处理,使得河流在洪泛期内的持水能力得到最大限度的提升。

植被设计对变化的水位线位置、当地的土壤种类、栖息地的多样性与景观的审美价值非常敏感。湿地空间由多种湿地草群组成,低矮灌木簇丛与沼泽长草族群伴生其中,并且在水位线升高时期,植被逐步过渡为周期性森林群落。河流湿地边缘是岛屿湿地生态系统的命脉所在。湿地梯田、斜面坡地、半岛与岛屿等生态板块都为湘江地区的多种野生物种提供了珍贵的栖息环境。湿地系统为候鸟、鱼类、两栖动物、有益昆虫提供了觅食、筑巢、育幼的栖息空间。

水生区域与滨河区域的植被设计取决于每个特定植物品种的水文适应能力。湿地区域遍布各种不同的植物,植被范围从挺水类的湿地草本到季节性洪涝期生长的灌木不等。湿地岛屿高地上的簇丛灌木区域可为部分野生物种提供栖息微环境。

湿地生机勃勃、形式多变,沿河高地与滨水低地的设计增强了岛屿生态系统的抵抗力。乡土滨河喜湿树种构成了低地森林的骨干树种,例如水杉与沙柳,均是中国常见的河道树种。

Project Purpose

The Project hopes we hope to emulate the strength and resilience of this river ecology, designing not only a large-scale public park, but an ecology that is open to change and rebirth.

Baxi island is one of the three islands in Xiang River adjacent to the city centre of Changsha. The name of "Baxi" means "Sand bar"; a fitting description for the islands. River forces have shaped the islands, under varying conditions. A wider river channel and higher river velocity have led to river bank erosion at its east edge, while a narrow water channel and lower river velocity has led to the establishaent of natural wetland at the west edge. In nature, river island land form follows the river flow direction and movement. Interior channels are formed as a result of seasonal flooding, and island vegetation patterns reflect soil and elevation variations.

Designed to accept seasonal river level fluctuation, Baxi Island is shaped into a series of linear berms and terraces. The directional nature of the landform allows for rising waters to flow thru and over the island. The sides of the island are stepped to create wetlands planted with various species, and act as a protection buffer against river currents. The meandering wetland terrace shape also celebrates the local picturesque rice paddy

设计策略

顺应河流与岛屿生态的竖向设计
将岛上的建筑、构筑物和步行道布局在最高水位线以上,以抵御季节性洪水的侵袭,这就需要额外的土方。在塑造地形起伏的同时,最小化岛上的填土和挖方,使河流在洪水期有足够的径流量通过区。

多样的水岸形式
在水面和湿地下加建结构性堤围,固定已经受到严重侵蚀的河岸。
西侧水岸河水流速较慢,因此可以设计更多在植物中游乐漫步的自然湿地水岸景观。

建立植物群落和栖息地
竖向高度不同、季节性洪水的影响,多样性的湿地、低地、中等高度和高地植物会在不同区域形成茂密的植物区域。这给多种动物在岛上的湿地、台地、河谷和池塘等不同生境中停留和栖居都创造了条件。

融合多种活动与功能
步行道和花园形成的游人活动网络遍布全岛。步行道会给不同需求的散步者或者徒步爱好者提供不同等级的环岛路线。公园、广场、户外剧场、眺望台和野生动物观察点分布岛上,让游人的旅行变得妙趣横生又寓教于乐。

landscape. At higher elevations, the wetlands give way to bermed forest planted with majority of native trees and grasses. Higher terraces are created for meeting facilities, pavilions, and museum. The south wetland area contains a blend of wetland islands and small wetland peninsulas that are linked by a meandering boardwalk network. Their elevation is slightly lower than the island to allow seasonal flooding of the marshes during the high water season and support various emergent wetland plants.

While the wetland terraces frame the island and provide a level of erosion protection, trails and boardwalks link a series of pavilions, overlook piers, ampitheater, marina, wetland museum, sculpture gardens, and meeting villas beneath a forest of planted native Metasequoia.

Discovery Gardens

A planted native Metasequoia forest provides the backbone of Changsha Island, from which a series of gardens and meadows emerge. An extensive trail, boardwalk, and bike path system provides access to the island's wide variety of discovery activities. "Dragon Gardens" recall the rich history of Changsha, one a stone mosaic anchored by water springs bubbling between large stones. Another "Sky Dragon" is created from fabric and steel flying thru the trees. A large bamboo garden hides a secret tea house, perched over a water lily pond carved from a forest clearing. The Children's Garden is anchored by a large tree house, rope swings, and slides. A series of ponds provide additional habitat for smaller aquatic species. The Changsha Baxi River Interpretation Centre tells the story of an ancient river city that became the trading gateway into central China, and at one time the capitol of the "Middle Kingdom". Miles of trails provide access to quiet meadows and shaded forest, a perfect sanctuary for meditation from the prosperous City of Changsha.

At Changsha Baxi Island, both the forces and science of nature are blended into an artful experience, creating a new river oasis in the heart of the city.

Stabilizing the edge

The east edge of the island parallels high velocity river flows, where erosion has been a consistent condition. To stabilize this bank, a submerged series of engineered embankments are built, allowing for the establishaent of wetland terraces at the waterline. A continuous strip of aquatic shrubs and deeper rooting wetland plants, such as phragmites and typha species, serves as the green screen along the east bank where erosive forces are the greatest. This linear network of grasses will help to slow water velocities during high water periods, providing stream bank protection along the east side of the island. The vegetative bank also functions as a habit opportunity for birds and wildlife alike.

The west edge is influenced by lower river velocities and higher degrees of sedimentation. Thus, the existing condition is dominated by natural wetlands. These are preserved and extended, with additional meandering wetland terraces and slopes created at varying elevations per the seasonal flooding regime.

River systems like the Xiang River historically had large quantities of wood along its river bank. This wood was the result of large floods that would take down trees and deposit the material on islands and stream banks. Wood in river systems plays an important role in stabilizing the river bank and providing important habitat for wildlife. The southern wetland provides an important opportunity to integrate large woody debris into the restoration design providing both structure and habitat. Several engineered log jams will be placed at the beginning of the grass-lined south wetland channels (in the east) to direct water into the channels and away from the river bank to limit scour. These structures dissipate the large forces that flood waters cause, protecting the structural integrity of the island and the diverse wetland grass and shrub vegetation communities.

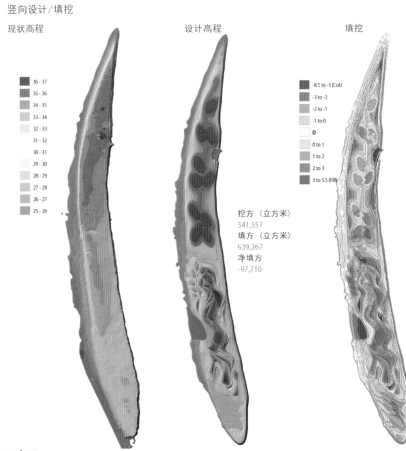

顺应河流与岛屿生态的竖向设计
vertical design on the foundation of the river flow and island ecosystems

Water Fluctuation

The river fluctuates annually from elevation 30 m to 36 m. Berms and terraces are designed to allow for water to move through and across the island, and celebrate the seasonal changes. The landscape is designed to accommodate these fluctuations with different plant communities occurring at different elevations.

Based upon existing island topography, the grading design minimizes cut and fill across the length of the island, maximizing the river carrying capacity at flood season.

The planting design is sensitive to fluctuating water levels, local soil types, habitat diversity, and aesthetic landscape value. The wetland area consists of multiple wetland grass communities intermingled with clusters of low shrub and marsh zones, and transitioning to a seasonal forest at the upper elevations. The river wetland edges serves as the lifeblood of the island wetland eco-system. A mosaic of wetland terrace, slope, peninsulas and islands offer valuable habitats for various wildlife species native to the Xiang River region. This wetland network provides foraging, nesting, and rearing habitat for migratory birds, fish, amphibians, and beneficial insects.

Aquatic and riparian area planting designs are based on the hydrologic adaptability of each specific plant species. A variety of plant species encompass the wetland zone, ranging from emergent wetland grasses to seasonally flooded shrubs. Clustered shrub areas are designed at the high-point of several wetland islands to provide micro-habitats for wildlife species.

The dynamic variability and land forms of the wetland, riparian lowland and highland design increase the resilience of the ecosystems among the island. Native riparian trees form the backbone of the lowland forest, such as Metasequoia and Salix species, the tree common to the rivers and canals of China.

东岸
河流宽阔，水流速度快，造成大量水土流失。

西岸
河流狭窄，水流速度慢，造成泥沙淤积，形成天然湿地。

巴溪洲现有岸线
Current Brazil Island shoreline

洪水流向分析

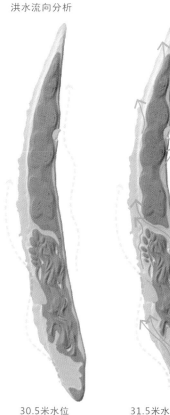  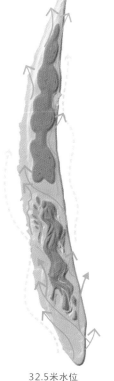 

30.5米水位　　　31.5米水位　　　32.5米水位　　　34.5米水位

30.5 - 37　　　31.5 - 37　　　32.5 - 37　　　34.5 - 37
27 - 30.5　　　27 - 31.5　　　27 - 32.5　　　27 - 34.5
约300天/年　　约50天/年　　　约15天/年　　　约2天/年

缓冲水位波动
景观的设计是通过不同高度的不同植物群落来缓冲消化季节性洪水对岛屿的影响，护堤的设计也允许水流穿过岛屿。

灵活的竖向设计
灵活的竖向设计使得在保证河流在洪水期的流量规模下，整个岛上的填土和开挖都保持最小。

顺应河流与岛屿生态的竖向设计
Vertical design on the foundation of the river flow and island ecosystems

软质//西岸
水流：慢
边界：无堤围
边界形态：曲线
植物种类：镶嵌湿地植物

硬质//东岸
水流：快
边界：水面下工程堤岸
边界形态：相对平直
植物种类：深根湿地植物

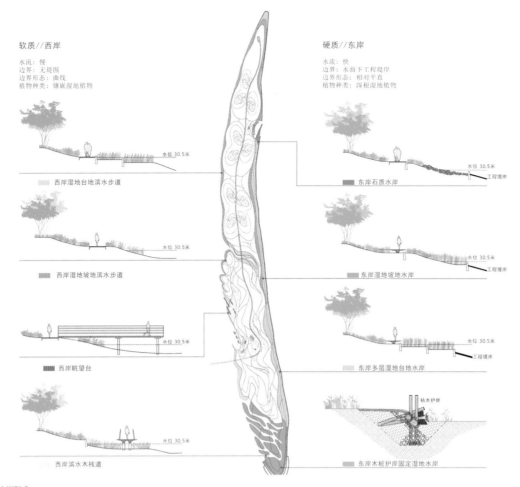

多样的水岸形式
Diverse waterfront forms

通过高程定义植物群落

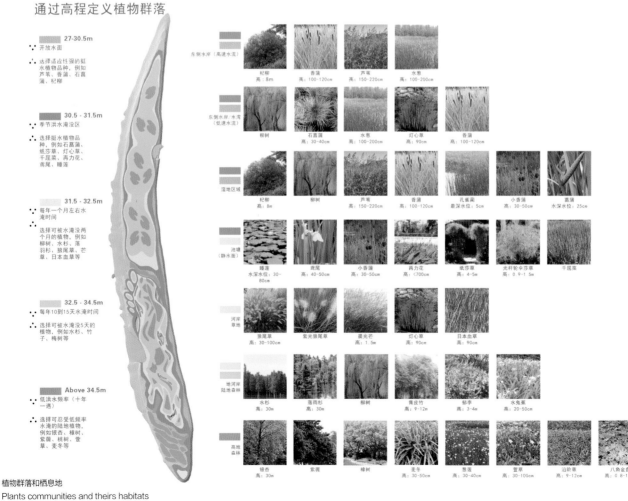

植物群落和栖息地
Plants communities and theirs habitats

冬季本地芦苇
Local reed in winter

春季油菜花与竹林
Canola flower and bamboo forest in spring

南部湿地（冬季）
Southern wetlands (winter)

南部湿地（鸟栖木）
Southern wetlands (bird perches)

南部湿地 (夏季)
Southern wetlands (summer)

水杉林 (夏季)
Metasequoia Forest (summer)

196

天津文化中心生态排水防涝系统设计

ECOLOGICAL DRAINAGE & WATERLOGGING PREVENTING SYSTEM IN CULTURAL CENTER, TIANJIN

项目位置：中国天津市
项目规模：90 ha
设计公司：安博·戴水道
翻　　译：张旭

Location: Tianjin, China
Size: 90 ha
Design Firm: RAMBOLL STUDIO DREISEITL
Translated by Zhang Xu

项目概况

天津位于中纬度亚欧大陆东岸,北温带东亚季风盛行的地区,位属暖温带半湿润大陆季风型气候。天津文化中心位于天津河西区,西侧毗邻迎宾馆和天津大礼堂,总面积约90 ha。

项目所在地原为天津乐园,为市中心较为开放的公共区域,周边市政雨水管网排水能力低于一年一遇水平。然而,由于原有大面积透水性较好的绿地和水面,径流产量较小,雨水基本不外排。

开发后场地中约56万m^2为不透水新建区,包括43万m^2硬质屋顶和铺装,如果不采用任何雨洪管理措施,势必产生成倍洪峰,严重威胁城市核心地带的安全。基于提升雨洪应对能力的重要性和迫切性,文化中心的雨水排水系统设计标准需要合理提高。在这种形势下,一套具有雨洪收集、预处理净化、调蓄和强排结合的综合排水防涝系统经过多轮论证后被采纳。

设计理念:海绵城市

传统的雨水排放方式无法灵活应对地块开发后径流增加的问题,原因是径流扩大量往往会大大增加雨水管网及配套设施的规模和造价,破坏自然水循环模式后容易造成内涝,还有附带的面源污染问题。该项目创造性地提出"海绵城市"的设计理念,即将城市雨洪管理与城市空间开发相结合,运用先进的雨水收集技术,场地内大部分暴雨径流经雨水收集系统收集、过滤、流入中心湖后进行存储再利用并循环净化。同时,得益于调蓄沟的削峰功能,场地内的洪涝风险得到有效控制。创新性水敏城市设计作为该工程设计的重中之重,力求实现以下几个目标。

(1)提高排水防涝的标准。
(2)在高密度的城市中心区域改善自然条件和微气候。
(3)节约市政排水基础设施的投资和维护费用。
(4)作为生态基础设施融入城市发展之中。
(5)符合生态排水防涝设计标准。
(6)按照雨水管渠设计重现期采用特大城市中心城区3~5年的标准,雨水调蓄排放标准为三年一遇,降雨历时两个小时;开发后的外排径流峰值不超过开发前的径流峰值,中心湖体最高水位和常水位之间的雨水调蓄排放标准为百年一遇且不外溢。

概念设计流程

降雨发生后,内排区域屋顶、道路和绿地的雨水将被不同形式的排水沟、入流井和雨水管道汇集在一起,经过沉淀井的预过滤和沉淀,进入地下模块化调蓄沟里。在调蓄沟中,水流得到滞留,雨水峰值变小,并通过终端设置的泵房流入生态净化群落,经过净化过后进入中心湖中储存。外排区域的雨水也会经过收集和沉淀的过程,最后通过调蓄沟滞留后流入市政管网。

1. 场地排水分区划定

根据场地整体布局和竖向规划,场地可划分为22个排水分区流域。其中内排(排入调蓄沟和中心湖后再溢流入市政管网)和外排(排入调蓄沟后进入市政管网)的面积分别占场地总面积的53%和47%。

2. 管网和设施基本布局

经过评估和测算,各地区的雨水收集系统主要包括以下几个部分。

缝隙式排水沟:V形横截面缝隙式排水沟和盖板排水沟采用树脂混凝土成品排水沟,可高效地收集雨水。

沉淀井:通常状况下,雨水初期径流夹杂杂质及异物,无法达到进入泵井、管道和湖体要求的纯净度,预处理是非常必要的。

调蓄沟:调蓄沟的主体为具有骨架结构的"蓄水模块箱",其由高强度的合成物质制成,具有很强的承压能力和超过传统蓄水设施的空间利用率。

中心湖:中心湖面积97 900 m², 设计常水位2.2 m², 三年一遇的最高水位2.5 m, 最低水位2.1 m。

通过调蓄沟和中心湖的串联调蓄,三年一遇暴雨的情况下,雨水外排的量由14.2 m³/s减少至0.75 m³/s;百年一遇的情况下,依靠中心湖和岸顶标高之间的浮动空间(0.9 m),亦可保证降雨不外溢,从而保证场地在极端降雨情况下的防涝安全。为了维护中心湖的水质,水体通过自然净化群落以400 m³/h的流量进行循环净化,长期达到地表水III类水质标准。

实施与运营

天津文化中心生态排水防涝工程中明确了排水防涝的标准和功能,对于雨水收集、净化和再利用的问题进行了多方面的设计。所有不透水表面上的雨水都经雨水收集系统被收集和预处理,在入湖流程中由沉淀井等设备过滤,由模块化调蓄沟调蓄,最后经生态群落净化之后被储存起来进行再利用。经过一段时间运行,证明此套系统为城市带来了良好的生态效益和经济效益,也对国内其他类似的市政排水和雨洪利用工程具有一定的参考与借鉴作用。

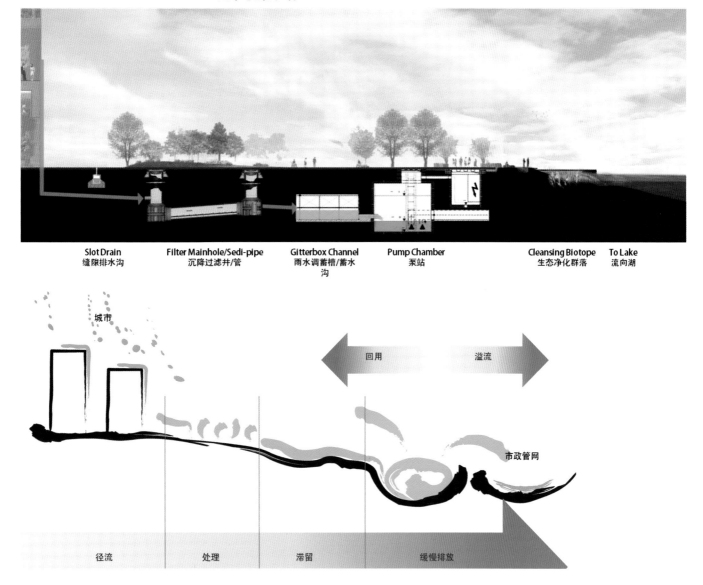

General Situation

Tianjin is located in the mid-latitude region, at the east coast of the Eurasian continent, north temperate regions, East Asian monsoon prevalent area, and belong to the warm temperature & semi-humid continental monsoon climate zone. Tianjin Cultural Center is located in Hexi District with a total area of approximately 90 hectares. Its west side is adjacent to the Guest Reception Hotel and Tianjin Great Auditorium and.

Formerly, this site is a park, a comparatively exoteric public area in the central city. The standard of the flood drainage & municipal sewer network in the surrounding parcels was under once a year flood level. However, due to a large area of original green space with better permeability, the runoff yield from the site doesn't need the extra space to be discharged.

After the construction of 560,000 square meters impervious area, including 430,000 square meters roofs and pavements. The floods peaks are bound to doubled or even tripled, and would impose a severe threat on the security of urban core, unless to implement the stormwater management to address the critical issue. It is crucial and pressing to improve the resilience and flexibility when confront with the storms and floods, so the design standards of the Culture Center drainage system should be upgraded in an appropriate range. In this situation, a set of waterlogging-preventing drainage system synthetized with the functions of storm water collecting, pretreatment purification, water regulating through retention & detention process and potent emission ability, has been adopted after several rounds of argumentation.

Design Philosophy: Sponge City

The traditional drainage methods cannot respond to the increased runoff flexibly after the developed land has been expanded or enlarged. The increased amount of runoff tends to proliferate the scale and cost of the storm sewer pipe network and the accessorial facilities. The disturbance to the natural water circulation is waterlogging prone, and is likely to trigger the interrelated non-point source pollution. The project put forward the idea of sponge city design innovatively, literally it is a solution plan in which combines the storm water management and urban space development together. Employing the advanced concept of rainwater harvesting, most of the stormwater runoff on site would be collected, filtered, circulated by the rain harvesting system, then discharged into the central lake to be storage, reused and purified. Due to the flood peak clipping function of water-regulating ditch, the risk of flood and waterlogging has been effectively controlled within the site. The idea of Innovative Water Sensitive Urban Design as one of the most significant engineering design philosophy embodies the following comprehensive objectives.

(1) Upgrade the standards of stormwater drainage and waterlogging preventing

(2) Improve natural conditions and microclimate in urban centers with high density

(3) Economize the investment and maintenance expense on the municipal drainage infrastructure

(4) Incorporated into the urban development as ecological infrastructure

(5) Ecological drainage and waterlogging preventing design standards

(6) According to the recurrence interval requirement of the stormwater pipes & channels design, the once 3~5 years standard in central megacity is adopted. The regulated rainwater emission standards is that, in the circumstance of once 3 years rain, 2 hour consistent precipitation won't result in the climax of runoff peak surpass the one before the site was developed, and in the terms of Cuiming Lake, space between highest water level and constant water level could contain the volume of the once 100 years precipitation without overflow and efflux.

Conceptual Design Process

After the rainfall occurs, the runoff from roofs, roads and green spaces in the internal discharge area will be delivered to sewer ditches, inflow wells, rain pipes & channels in different forms. After the pre-filtration and sedimentation process in the sedimentation wells, then the runoff will be

discharged into the underground modular water-regulating ditch. In the water-regulating ditch, rainwater will be retained and detained, the runoff peak declined, the terminal-installed pump room will sent the water to eco-purification vegetation communities, and eventually flow into the Central Lake for storage. As far as the runoff in the external discharge area, the process of harvesting and sedimentation is inevitable as well, and finally conveyed by the water-regulating ditch to the municipal sewer network after experienced the process of retention and detention.

1. Drainage zoning on the site

According to the overall layout and vertical design of the site, the entire place can be divided into 22 drainage basin zone. The internal discharge area and the external discharge area respectively cover the whole area in proportion of 53% and 47%.

2. The basic layout of the pipe network and facilities

After calculation and analysis of each zone, Its system mainly consists of several parts.

Gap gutter: The split gutter of V-shaped cross-section and the plank-covered gutter made of resin concrete could help to collect and harvest rainwater efficiently.

Sedimentation well: usually, there are lots of impurities and particles mixed in the original stormwater. The runoff couldn't discharged into the pump well, pipe channel, and the lakes directly, because the water quality of the runoff is far beneath the minimum purification requirement. In this circumstance, pretreatment to the runoff is very necessary.

Water-regulating ditch: The principal part of the regulating process is based on the module storage box with framed structure, which is made of high-strength synthetic materials. They have a strong bearing capacity and embraces a relatively higher space utilization rate comparing with the traditional storage facilities

Center Lake: the area of the Central Lake is 97,900 square meter, the designed constant water level is 2.2 m, the highest water level and lowest water level in the case of once 3 years rainfall is respectively 2.5 m and 2.1 m.

Through a series connection of water regulating ditches and the Central Lake, when encountered with the once 3 years precipitation, the amount of rain efflux would reduced from 14.2 m^3 / s to 0.75 m^3 / s. In the extreme condition of once a 100 years precipitation, the flexible space between the Central Lake and the lake shore top elevation (0.9 m) could hold the rainfall to avoid the overflow and efflux, in this way to prevent waterlogging and guarantee the site security. To maintain the water quality in the central lake, water would flow through natural treatment community in the speed of 400 m^3 / h and experienced the process of circulation and purification in expecting to achieve class III water quality in a long-term.

Implementation and Operation

Ecological drainage & waterlogging preventing project of Tianjin Cultural Center identified and cleared the standards and functions of this kind of system, and focusing on the rainwater harvest, purification and reusing issue, a series of multi-purpose design were put forward. All the rain runoff on impervious surfaces is collected and pretreated by rainwater harvest system, then to be purified in sedimentation wells and be moderated in the water regulating ditch during the infill-lake process, and eventually flow into the ecological communities to be stored for reusing. After a period of operation, it has been proved that this system could bring the city with abundant ecological and economic effectiveness, and provide some valuable experiences for the similar municipal drainage program and rainwater utilization projects in domestic area and overseas.

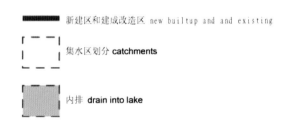

新建区和建成改造区 new builtup and and existing

集水区划分 catchments

内排 drain into lake

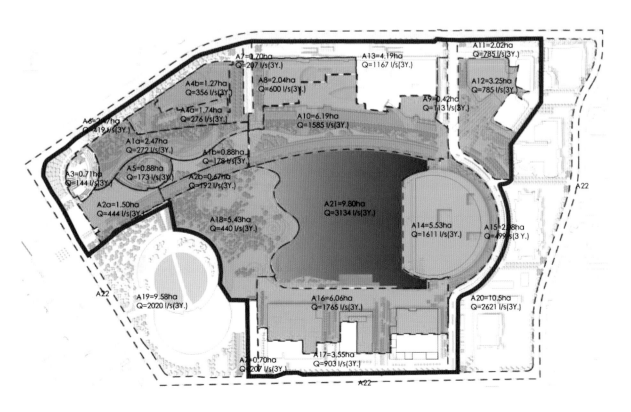

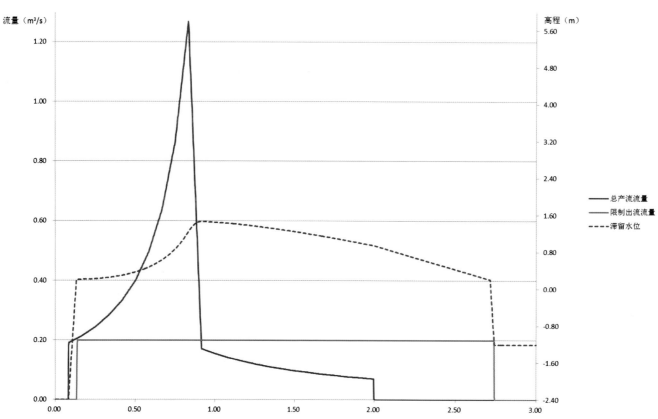

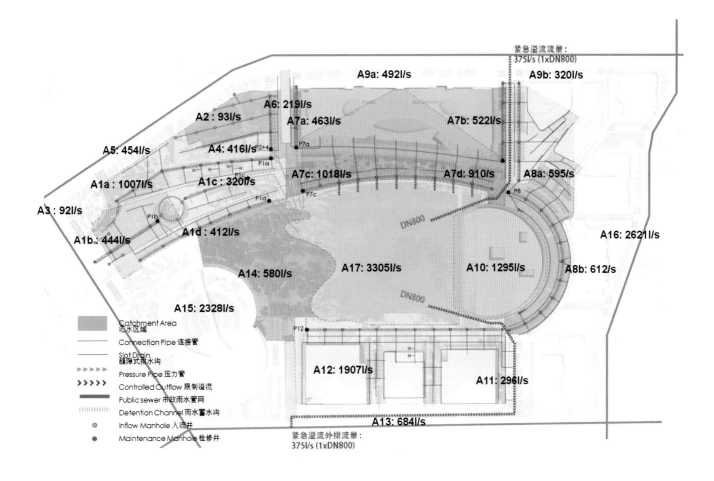

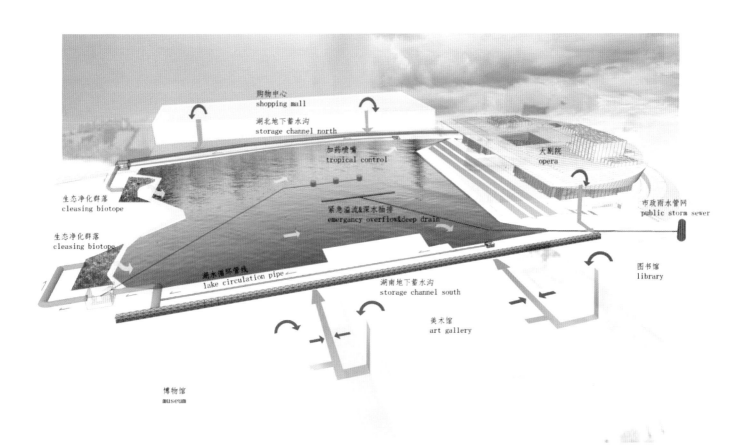

204

生态校园的综合设计理念与实践：辽宁公安司法管理干部学院新校区设计

CONCEPT AND PRACTICE OF THE ECOLOGICAL CAMPUS DESIGN: LIAONING POLICE AND JUDICIAL MANAGEMENT CADRES COLLEGE NEW CAMPUS DESIGN

项目位置：中国辽宁省沈阳市
项目规模：24.7 ha
设计公司：土人景观
翻　　译：张旭

Location: Shenyang, Liaoning Provice, China
Size: 24.7 ha
Design Firm: Turenscape
Translated by Zhang Xu

辽宁公安司法管理干部学院新校区位于沈阳市棋盘山风景区辖域内，沈棋路以南、沈抚公路以北，西有世界遗产清东陵，东有沈阳世博园，南临沈阳的母亲河——浑河。规划用地面积24.7 ha。场地总体态势是三山环抱，北高南低。场地中部地势最为平缓，呈现以缓坡和台地为主，逐渐向南部入口区下降的态势。场地景观特征为"两山含一谷"。场地中央的谷地地势较为平缓，其东、北和西侧为山体所围，南眺浑河，背山面水，是典型的"风水宝地"。

挑战与目标

该项目在景观生态设计方面主要面临以下几个挑战。

（1）场地诉求：所在区域毗邻世界遗产地，是沈阳市的著名风景区和游憩地。物质上和视觉上对周围环境的消极影响都是不能被接受的。主管部门规定，该区域内的建筑高度控制在18 m（现状学生宿舍高21 m），绿地率需达到60%。如何使新校园成为风景中的建筑，且让校园环境与周边景观相得益彰，便成为该设计的最大挑战。

（2）绿色校园诉求：如何在优美的环境中建造一个全方位的绿色校园，体现当代的环境与生态理念，并彰显校园的个性与特色，成为衡量校园成功与否的重要标准。

面对上述挑战，设计师将校园与周边环境、功能与绿色诉求相结合，提出了"全方位生态校园"的设计理念，并获得了甲方的认同，使项目得以全面实施。

设计方案

该项目规划旨在营造一个人与环境和谐相处的大学校园，一个富有现代气息的大学校园，一个体现地方文化、自身特点和环境特色的大学校园，一个经济实用的大学校园。该项目规划遵循以下原则：设计遵从自然特点，发于自然，融于自然。布局分为"化零为整"的相对集中式建筑布局和院落式空间布局。建筑朝向东西，减小南北视线通廊上的视觉障碍。半覆土的建筑形式，建筑与地形的有机结合，建筑融入于自然。架空的建筑形式，减小建筑对植被的破坏。具体措施包括以下几个方面。

（1）制定整体性的生态规划。规划之初，明确场中作为绿色校园永久保留和可成为校园绿地和场所的景观，对场地内的自然与文化遗产给予充分的尊重。

（2）遵从自然地形，强调绿色体验的功能布局。规划结构为"三带两区一中心"。"三带"为生态景观带、教学办公带和生活带，完全结合地形而规划。"两区"为体育运动区和培训服务区，位于场地西部，相对独立。"一中心"指学院中心广场区。校园的最大特色是以中部的生态景观带为纽带，分割教学办公带和生活带。自然景观带设计充分利用场地中部原有冲沟，作为雨水收集和中水过滤净化的设施，利用地形，使其形成逐渐跌落的水体景观，既可改善校区小气候，又可为广大师生提供亲近自然的景观体验，位于学生每天来回于教室和餐厅及宿舍之间的必经之路上。

校园设计方案阶段鸟瞰图
Bird view of the campus in the design phase

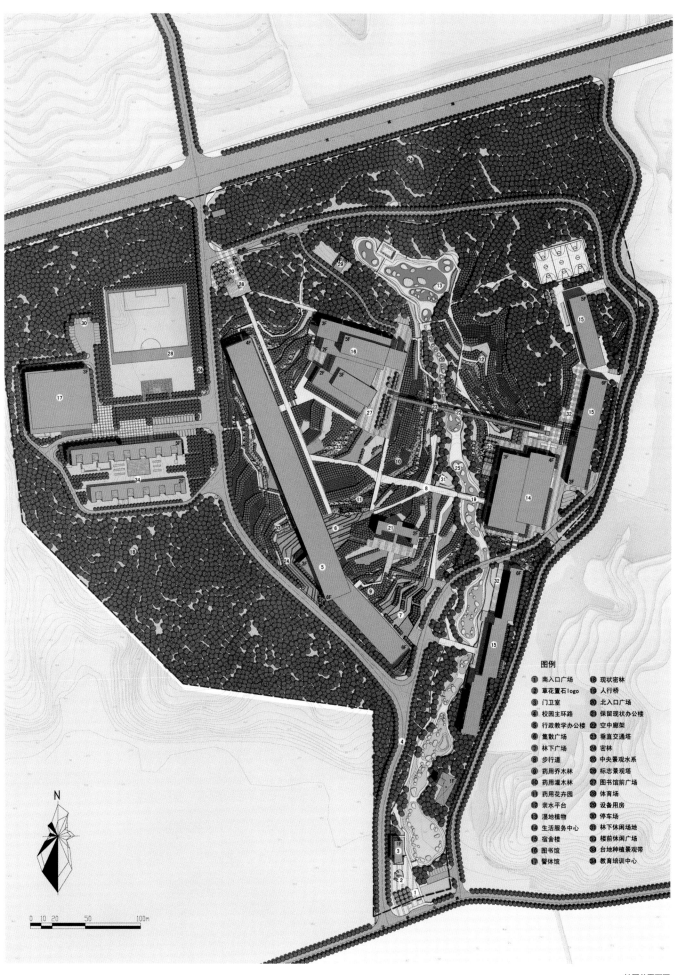

校园总平面图
Master Plan of the campus

（3）打造系统的校园景观生态设计。校园景观的生态设计集中体现在对原场地地形、植被和文化景观的保护和利用，雨水收集和利用，中水净化和循环利用，低维护植被的设计等各方面。

场地中原有的一条冲沟的地形和自然植被得到了完整地保留，并作为整个校园的主要雨水汇集区。沟的上下游之间有35 m的高差，通过生态化的堤堰，逐级跌落，形成多级湿地，调节雨洪，营造丰富多样的湿生栖息地。这个长达600 m的多级湿地系统同时被用于中水的进一步净化，每天有400~500吨的水量经过沿沟分布的系列湿地。沟谷两侧设计游憩木栈道、座椅和读书台。与学生中心相结合，设计了一直延至谷底水面的休憩廊桥和平台。昔日的黄土荒沟，而今成为校园最亮丽的风景带。

为减少维护建设的成本，校园绿化设计大量使用乡土植被。大量的野花组合布满山坡和林下，与当代风格的建筑相得益彰；乡土灌木用于护坡、建筑边沿的基础种植和沿水沟的蔓延，形成生机勃勃的拟自然植被景观。在保留原有树木的基础上，补种本地乔木，形成森林景观。

结论

从2006年12月开始，从概念设计到基本建成，用了整整5年的时间，甲方与设计方紧密合作，坚持不懈地探索绿色校园的实现途径，提出并践行生态校园的综合设计理念，从自然景观的保护盒利用，到乡土植被和低维护绿化的广泛采用；从雨洪的生态利用，到污水的生物净化和中水回用；从楼顶的太阳能热水，到室内免冲便池的使用；从建筑设计中自然光的利用和建筑节能考虑，到步行校园理念的实现，生态校园的综合设计理念已得到了全面的实践。总结起来，其特色可概括为以下几个方面。

第一大特色：一岭探碧水，二龙戏朱雀，苍崖半环山，汇流竟成趣。

第二大特色：层绿依山形，台地师自然，回路外作环，筑屋互成院。

第三大特色：近水听笑语，跨溪闻书声，柔情可比水，刚直自有形。

第四大特色：南眺尽双眸，北望隐深林，内能怀山水，外可安四方。

（本文节选自：俞孔坚,张慧勇,文航舰. 生态校园的综合设计理念与实践:辽宁公安司法管理干部学院新校区设计[J]. 建筑学报,2012,03:13-19.）

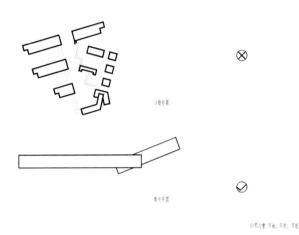

归零为整（节地、节材、节能）
Gathering parts into a whole (land-saving, materials-saving, energy-saving)

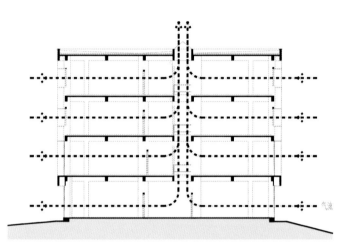

自然通风设备
Natural ventilation equipment

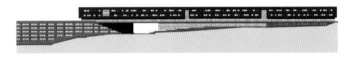

教学楼
Teaching building

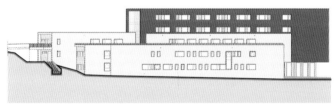

图书馆
Library

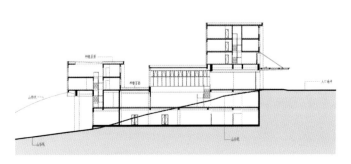

建筑布置于山谷中，缩减体量，冬暖夏凉
Buildings in the valley, body reduced and mass decreased, cool in summer and warm in winter

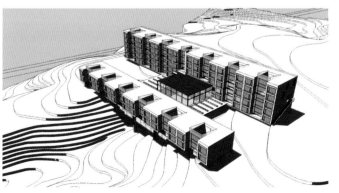

培训中心
Training center – buildings layout in the valley

| 源头汇水区 | 浅水湿地过滤区 | 深谷集水区 |

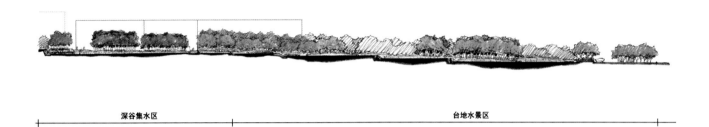

| 深谷集水区 | 台地水景区 |

校园设计方案阶段模型
Model of the campus in the design phase

The new campus of Liaoning Police and Judicial Management Cadres College is located in the administer region of Qipanshan scenic area, south to the Shenqi Road, north to the Shenfu Road, the World Heritage Eastern Qing Mausoleums on its west and the Shenyang Expo on its East, and Hun River, the mother river of Shenyang, flowing quietly on the southern side of the campus. The planning area of campus is about 24.7 hectares. The overall typology of the site is ringed on three sides by mountains, north high and south low. The terrain of the central site slopes gently, dominant by flat gradient and bench terrace, and gradually downward to the southern entrance area. The landscape of the site is featured of One Valley Embraced by Two Mountains. The terrain of the central valley is relatively flat, and mountains surrounded by its eastern, northern and western sides, water on its front and hills on its back, literally south to the Hun River, a typical blessed land according to the standards in Chinese FENGSHUI.

Challenges and Objectives

The major challenges in the ecological landscape design of the project are as follows.

(1) Site demands: the site is adjacent to World Heritage Area, a well-known scenic spot and recreational space in Shenyang. The negative impact on the surrounding environment is not acceptable no matter visually or physically. According to the administrative provision, the height control of the buildings is within 18 meters in this region (the student dormitories are 21 meter high), besides, the green rate should reach 60%. The greatest design challenge is to blur the boundary between the architecture and scenery, and bring out the best in each other.

2) Green campus pursuit: How to create a full-round comprehensive green campus in the such an beautiful environment to demonstrate the contemporary concept of E2 (environment & ecology), and to delivery the characteristic and individuality of the specific campus, are the standards to measure whether campus construction is successful or not.

Confronted with these challenges, the design team integrated the campus to the surrounding environment, combined the building function with the demands for the green space. Meanwhile, they proposed the full-round comprehensive green campus design concept and gained the recognition from the consignor. In this circumstance, the concept of the project has been fully implemented.

Design Plan

The design aims at creating a university campus where the human beings inhabit harmoniously with nature, where permeates with a sense of modern & contemporary, where boasts of the local culture, individual characteristic and the environmental features, where based on the principle of economy & pragmatism. During the planning process, the design team followed the idea of complying with nature, inspiring by nature and integrating in nature. In line with the rule of assembling the parts into a whole, composing the buildings in a relatively centralized layout and a courtyard-like space style. Place the building towards east-west orientation, in order to reduce the visual obstruction in the sight line of north-south direction. The architecture of the semi covered soil styles, organic union between architecture and landform, buildings melted in nature, overhead construction forms, all these contribute to the decrease of the buildings' negative impact on vegetation. Other details include several parts.

(1) The master ecological planning. At the initiative stage of the planning, it

is clear that the space for the green campus and the landscape for the local place should be reserved permanently. And during the planning process, the respects towards nature and heritage on the site should be delivered and displayed adequately.

(2) Complying with the natural landform, emphasizing the functional layout in which could render people with green experience. The planning structure was Three Zones & Two Areas & One Center. The Three Zone refers to the ecological landscape zone, educational & office zone and residential zone. In the plan, the terrain and landform on the site has been fully taken into consideration. The Two Areas refers to the sports area and training & services area. They are located relatively independent on the western site. The One Center refers to the College central square area. One of the most obvious features on the campus, in which employed the ecological landscape belt as a linkage, dividing (also connecting) the educational & office zone and the residential zone. In the process of natural landscape belt plan, the design team took fully advantage of the existing gullies on the original central site, reforming them as the facilities for rainwater harvesting and reclaimed water filtration & purification. Utilizing the terrain and landform on the site, to form a stepwise falling water landscape. In this move, it not only can improve the microclimate on the campus, but also provide an intimate natural landscape experience for the majority of teachers and students. Besides, this is the only way between the restaurant and dormitories by which the teachers and students go back and forth every day.

(3) Ecological Campus Landscape Design:Ecological Landscape Design of the Campus focus on the protection and utilization of the original site topography, vegetation and cultural landscape, stormwater harvest & storage, reclaimed water recycle & reuse, low-maintenance vegetation, and other aspect.

The gullies in the original topography and the intact reserved natural vegetation were served as the main stormwater catchaent facilities on the site. The dispersion of the elevation between upstream and downstream is up to 35 meters. By formulating the ecological dikes, step by step, the water flow and drop, combined with land relief and transformed into the multi-leveled wetlands, helped to regulate and manage the stormwater, and to create a variety of wetland habitats. The 600 meters multi-leveled wetland system could also be used for the further purification of the reclaimed water. About 400-500 tons of water is delivered and distributed by a series of wetlands along the gullies every day. Recreational boardwalk path, chairs and reading desks are placed on the both sides of the gullies in the design. Combined with the student center, the entertainment bridges and platforms have been designed to rest on the water surface at the bottom of the gullies. The former deserted loessial gullies are transformed to the most brilliant scenery on a campus today.

In order to reduce the costs of construction and maintenance, the native vegetation and indigenous species were planted and seeded extensively on the campus. Abundant clusters of different wild flowers combination covered the hills and embellished the forests, the contemporary architecture and vegetation complement each other and both benefit by associating together. The native shrubs are mainly used for revetment and slope protection, basic plantation at the edge of the buildings and the plants extension & transition along gullies, these all formed a quasi-natural vegetation landscape. On the basis of the remaining trees on the original site, the indigenous arbors were replanted, forming a forest landscape.

Conclusion

Start in December 2006, it takes exactly five years from the phase of conceptual design to the phase of basically completion. The consignor cooperated closely with the design team during this period. They persistently explored the feasible ways and methods to achieve a green campus, proposed and practiced the concept of the ecological campus comprehensive design. They adopted the measures of natural landscape protection & utilization, native vegetation usage and low-cost maintenance landscaping in a large scale, namely, from the stormwater ecological utilization, to the biological sewage purification and reclaimed water recycle & reuse, from the solar hot water facilities on the roofs, to the rushing-free urinals & toilets indoors, from the consideration for the natural light usage & building energy conservation, to the implementation of the pedestrian-friendly campus concept. The idea of the ecological campus comprehensive design has been fully practiced. To sum up, its characteristics can be summarized as follows:

The first major feature: one ridge embodied by waters rippled with bluish waves, two dragons intertwined with phoenix covered by flaming feathers, verdant cliff forming a half-moon shape mountain, bountiful streams converging in an pleasant manner.

The second major feature: the layered greenery undulated with the mountain relief, the terrace construction inspired by the natural disposition, the roads link and alternate with each other to produce the contour lines, the houses assemble and allocate between one another to form the fold yards.

The third major feature: the laughter bursts overflowed from the riverside, the reading sound whispered beyond the creek streams, the tenderness is parallel to water, the integrity is everything but intangible.

The fourth major feature: take a panoramic view from the southern vantage point, the thick forest expend and emerged in the far northern skyline, an interior broad vision could hold countless mountains and rivers, an exterior ambitious mind could shoulder numerous duty and responsibility.

The artical is from: YuKongjian, ZhangHuiyong, WenHangjian. The concept and practice of the ecological campus comprehensive design, Liaoning Shenyang Management Cadres College new campus design [J] Architectural Journal, 2012,03: 13-19.

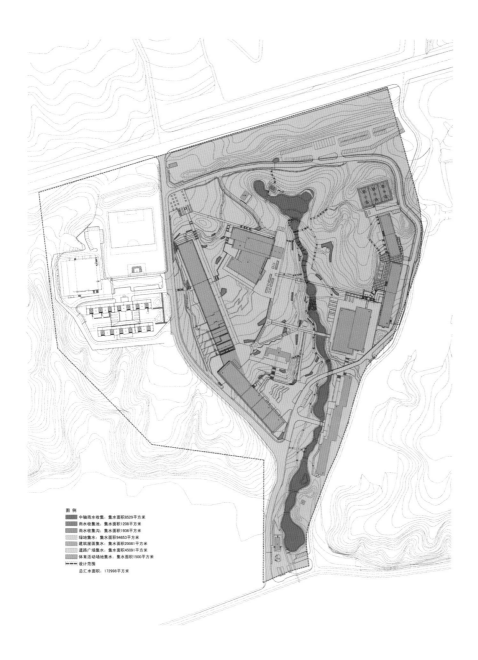

图 例
- 中轴雨水收集：集水面积8529平方米
- 雨水收集池：集水面积1208平方米
- 雨水收集沟：集水面积1938平方米
- 绿地集水：集水面积94653平方米
- 建筑屋面集水：集水面积20081平方米
- 道路广场集水：集水面积45091平方米
- 体育活动场地集水：集水面积1500平方米
- 设计水范围

总汇水面积：172998平方米

园内有多条木步道及多个休憩平台，使用废旧枕木制作，供休闲使用。
There are lots of wooden walkways and recreational platforms. They are made of old and waste sleepers. They are dedicated in the function of leisure and entertainment.

校园中心的雨水沟被改造成雨水收集和中水净化沟。
The gullies on the central site of the campus was transformed into the ditches for rainwater collection and reclaimed water purification.

学生中心前,一直延伸入谷地的木栈桥,利用经过湿地净化的水体,为学生营造良好的休憩环境。
In front of the student center, the wooden trestle bridge has been extended into the valley deeply, after flowing through the wetlands, the purified water-bodies provide the students with favorable creational environment.

214

引生境、承天露、生万物：
深圳湾科技生态园景观设计

CREATING HABITAT, COLLECTING RAINWATER, MAKING BIOLOGICAL REPRODUCTION : ECOLOGY PARK

项目位置：中国广东省深圳市

项目规模：20.3 ha

设计公司：深圳市北林苑景观及建筑规划设计院有限公司

Location: Shenzhen, Guangdong Province, China

Size: 20.3 ha

Design Firm: Shenzhen BLY Landscape & Architecture Planning & Design Institute Ltd.

深圳湾科技生态园坐落于南山区高新技术产业园南区，东临大沙河，毗邻深圳湾，处于咸淡水交汇处，是深圳水生态系统的重要组成部分。项目用地20.3 ha，其中建筑覆盖率约50%，是国家级低碳生态示范园区，实现绿色建筑全覆盖。

"智者乐水，仁者乐山"，设计在传承中国传统水文化和生态哲学观的基础上结合现代的生态理念与技术，通过"引生境、承天露、生万物"，立足于人与动植物生境以及水环境的营造，为居于现代高科技产业园的人们创造"知山知水"的生态美景，其中"仙人承露"更是对雨水的景观生态利用和精神内涵的双重升华。

该项目设计界面除地面景观空间，还涵盖地下空间、架空平台和屋顶花园(约17.6 ha)，同时引入"垂直城市"的设计理念，利用水的"流动和连通"，将收集的屋面、地表的雨水和生活污水通过雨水花园、生态草沟、湿地净化、透水铺装等方式，用于补充地下水和园区景观用水，经由水的流动循环，串联不同的景观，展现水幕、跌水、水阶、湿地等丰富的水态。

其中地下一层以多样绿岛加湿地花园营造配套商业，生态优美且拥有知识性、科普性的湿地环境。这个区域如同园区水处理的"心脏"，将收集到的雨水和生活污水通过湿地净化后，再回流入土壤。景观的打造模仿湿地自然形态，大量种植易维护管理的野生湿地植物，净化水质的同时，形成鸟类和爬行类动物的栖息场所，最终实现生态内涵与景观形式更全面的融合，构建建筑、景观、生态一体化的"全生命体"人居生态环境。

在景观艺术装置和细部处理上，该项目设计融入"海绵城市"的理念，打造既有独特艺术外形、又具生态内涵的景观地标。位于园区纵向景观轴线上的"生命塔"以红树中的木榄作为创作灵感，它有着优美的外形以及良好的生态价值，可固岸护堤，净化水源与空气，这正契合了深圳及这块设计场地的地域特色及人文精神。生命塔顶部树冠可以像花瓣般开合，树冠能承接上部雨水，下部雨水花园收集周边雨水，并将其集于中部雨水净化设施进行净化。净化后的雨水可用于浇灌立体绿化植物，顶部喷泉及产生喷雾用水。生命塔将技术和自然元素循环紧密关联并呈现出来，以求"引生境，承天露，生万物"的真正意境。"生长的红树"则选取深圳湾特有的红树枝叶形态进行抽象设计，结合新技术以及人的参与互动，使其在形成园区重要景观地标的同时，展示生态的设计理念对场所带来的美好景象。红树主体立面利用垂直绿化技术构造活的表皮，枝干以圆润饱满的形态出现，主要材料为钢架构及钢架膜结构的叶片组成。顶部"树冠"可以像花瓣般随天气变化进行开合，并承接雨水，用来塔内中心的景观水景以及垂直绿化的滴灌系统的补给。

该项目建成后将实现雨水花园4700 m^2，室外透水地面面积比率80.5%，场地径流系数0.6，并在全园实现节水灌溉技术的应用(整片绿地采用微喷灌，零散绿地采用滴灌技术)。该项目设计以节能环保技术为核心，引入"海绵城市"的设计理念，在打造国际一流低碳生态示范园的同时，必将成为高密度城市之中绿色海绵设计的典范之作。

雨水花园与中水花园技术

项目	雨水花园	中水花园
水源	天然雨水（活水）	人工处理后的水（死水）
种植基质	天然土壤	多孔骨料
结构	蓄水层、覆盖层、种植土层、砂层、砾石层	依赖物理、生物、化学手段及设备工艺
生态作用	涵养地下水	减少污水排放，水循环利用
物种丰富度	高	低
景观效果	自然、生态	人工痕迹重
运营成本	低	高

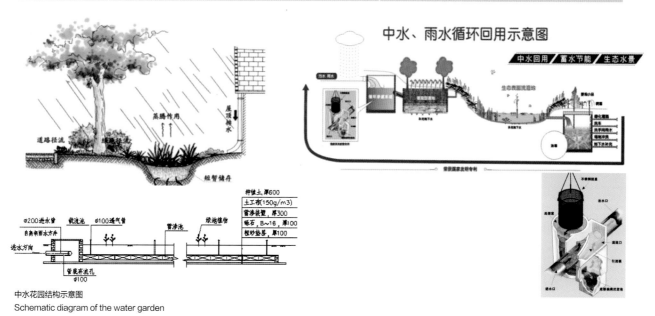

中水花园结构示意图
Schematic diagram of the water garden

Shenzhen Bay Technology & Ecology Park is located in the southern part of Nanshan High-Tech Park, adjacent to Shenzhen Bay and Dasha River, where in brackish intersection habitat. It is an important part of aquatic ecosystem in Shenzhen. Land use for this project is 20.3 hectares, and building covered about 50%. This park is also a national ecological demonstration zone of low carbon and green buildings.

"Intelligent people like water, merciful man like mountain", combined the modern ecology concepts and techniques with traditional Chinese culture and philosophy from the concept "Creating Habitat, Collecting Rainwater, Making Biological Reproduction". The design is based on the idea of creating habitats and water environment for animals and plant, creating ecological beauty for people who live in this modern high-tech industrial park. "Fairy collecting rainwater", this is double spirit sublimation to ecological use of rainwater.

The design boundary of this project covered 17.6 hectares, including ground level landscape, underground space, raised platforms, sky gardens and roof gardens. Designers refer to the design concept of vertical city, which is used water characters of "flowing and connecting", would collected roof rainwater, sewage through rain garden, ecological grass channel, wetland purification, permeable pavement. In order to supply groundwater and park water storage, demonstrating different series of water landscape, such as water curtain, drop water, water wetland and so on.

The underground level constructed commercial space with diverse green islands and wetland garden, built a beautiful ecological, knowledgeable and educational wetland environment. This area would collect rainwater and sewage through wetland purification and run back into the soil, like "water treatment heart". Landscape shaped through imitated wetland natural form and planted easy maintenance wild wetland plants. While it was purifying water quality at the same time, the space formatted habitats of birds and reptiles, which finally achieved integration of ecological intension and landscape forms. It constructed architecture, landscape and ecology into an "all living organism" a living environment.

Integrate the idea of sponge city into the landscape art facilities and detailed design, and then build landscape icons with both artistic features and ecological meanings. The creation idea of the Life Tower which is set in the north-south landscape axis of the park comes from the branches of mangroves. The mangrove with its marvelous appearance and beneficial ecological value not only strengthens the bank, it also purifies the water and air quality, and furthermore, it is perfectly coherent with the geological characteristics and cultural spirits of Shenzhen. The canopy of the life tower can be opened and closed like petals. The top canopy connects to rain garden above and the rain garden at the bottom to collect rainwater, which is purified in the middle section. Purified rainwater can be used to watering vertical greening plants. Fountain at the top provides spray water. Life Tower shows how to integrate technology with natural elements, to realize the idea of "Creating Habitat, Collecting Rainwater, and Making Biological Reproduction". The design concept of the "growing mangroves" is from the special shape of branches and leaves of mangrove. Use new technology and encourage public activities, then it becomes an important featured landscape construction displaying ecological approach of the design. Vertical greening is used here to create living surfaceand the main material of the leaves is steel structure and membrane. Canopy on the top can be opened and closed like petals when weather changes. It can collect water for irrigation of inside water landscape features and vertical gardens

After finished, the project will include 4,700 square meters of rain garden, the outdoor filtration pavement ratio achieves 80.5%, the Site runoff index reaches 0.6 and it applies the water-saving irrigation technology (the whole greening fields use microspray irrigation while the separated use drip irrigation). The project focuses on the energy-saving and environment-protection and it introduces the concept of sponge city. Finally, this international first-rate low-carbon ecological demonstration garden will become a model for building green sponge city in high-density cities.

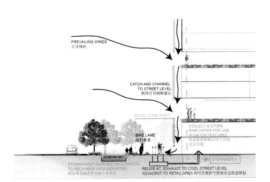

自然降温 Passive Cooling

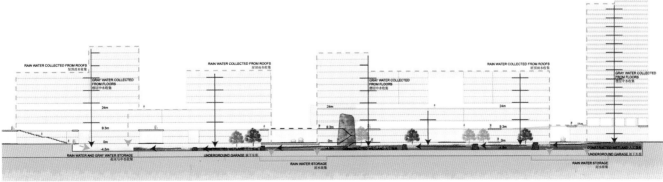

人工湿地 Constructed Wetland

人工湿地
Artificial wetland

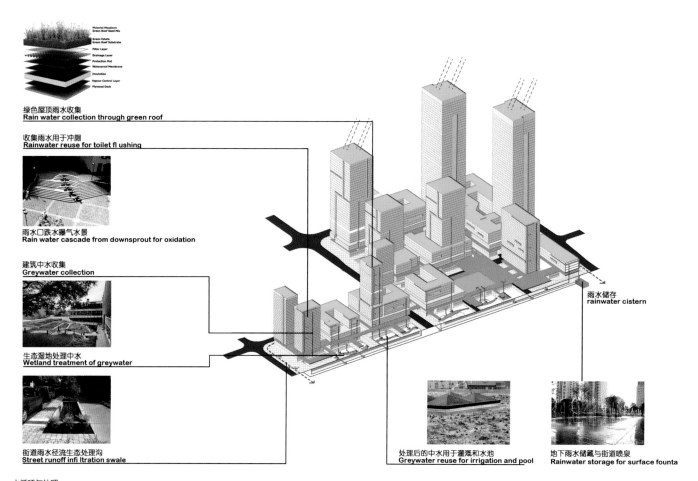

水循环与处理
Water recycling and treatment

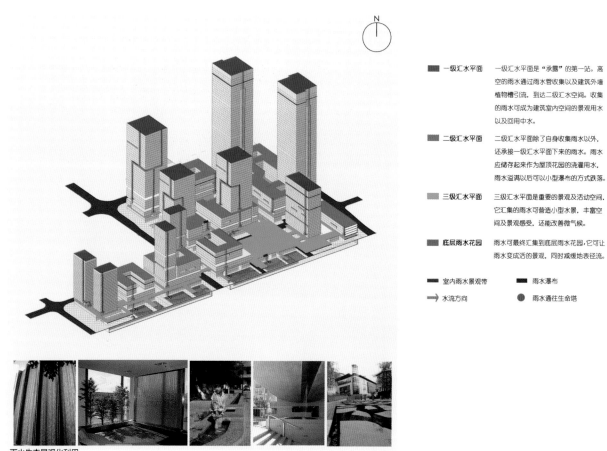

雨水生态景观化利用
Rainwater utilization in the ecological landscape

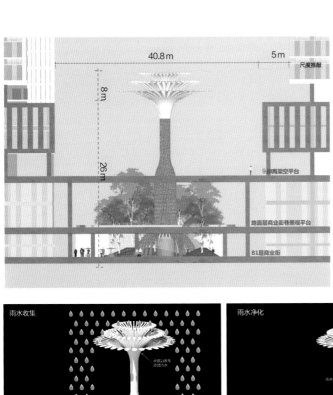
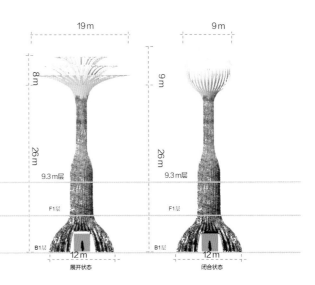
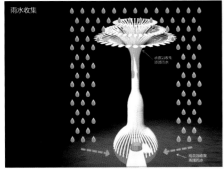
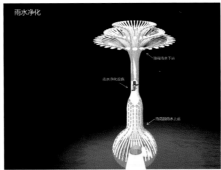
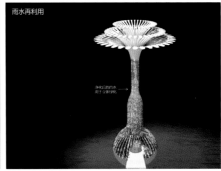
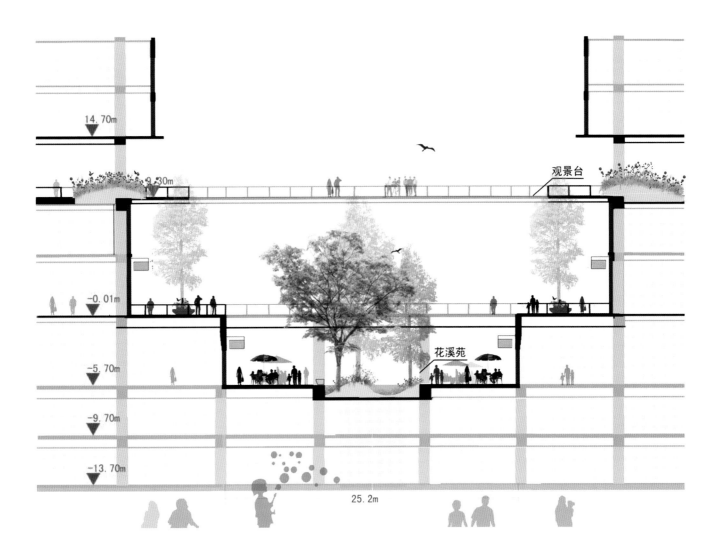

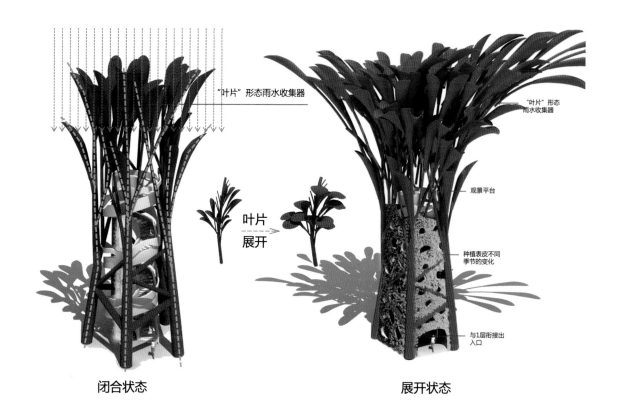

"叶片"形态雨水收集器

叶片
展开

"叶片"形态雨水收集器

观景平台

种植表皮不同季节的变化

与1层街接出入口

闭合状态　　　　　　　　　　　　展开状态

雨水花园
以沙河西路段为例

汇水范围

雨水花园

园区内雨水收集利用的两种形式：
1、入渗，补充地下水。
2、收集，净化后供园区内使用（浇灌、补充景观水等）。

 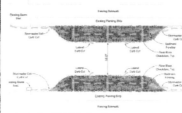

透水性路面铺装

定义：使雨水通过人工铺筑多孔性铺面，直接渗入路基，使水还原于地下的路面铺装。

适用范围：市政步道、园林甬路、树池、城市广场、体育场、住宅小区、停车场等。

社会效益：

1. 能够使雨水迅速地深入地表，还原成地下水，使地下水资源能得到及时补充，保持土壤湿度，改善城市地表植物和土壤微生物的生存条件来调整自然生态平衡。
2. 透水性路面具有较大的空隙率，并与土壤相通，能蓄积较多的热量，有利于调节城市大气的温度和湿度，减轻城市"热岛"效应。
3. 当下雨时，透水性路面能减轻道路排水设施的负担，减少峰值时段雨水管道向城市河流排放流量，透水性混凝土可防止路面积水和夜间反光现象，提高了人与车辆通行的舒适性和安全性。
4. 采用矿山废弃的石屑石粉和建筑垃圾制成透水地砖，还可以消化固体废弃物，更具环保理念。

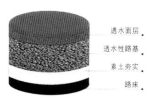

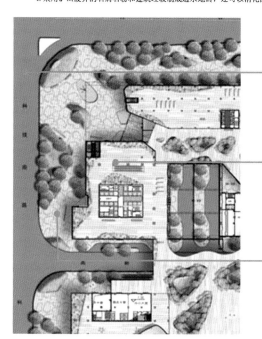

生态透水景观材料

以进口改性树脂为粘接剂，各种天然及再生材料（天然石子、树皮、炉渣、废橡胶、陶瓷碎片、碎玻璃等）为骨材，加入多种高分子添加剂，经混拌制成。主要用于透水透气性路面铺装，使城市土壤与大气的水、气、热交换生态体系得到改善。

非连续性拼接铺装

用实心砖，但砖与砖之间留出一定空隙，使天然的草能生长出来，增加地面的绿化面积。或空隙中填自然卵石，可以起到透水的作用，这种铺装方式可以运用在滨水绿地中。

生态透水砖

生态透水砖采用各种矿渣废料、陶瓷肥料、玻璃废料、石英砂等为主原料，经预制而成的再利用完全环保产品。由于其原料廉价，具有良好的透水、透气等功能，且变废为宝，因而具有良好的环保效益，在S120省道上使用，能起到很好的示范作用。

植草砖

连续性的钢筋混凝土石墩所构成，表面承重物或重车碾压后，不会造成不平均的沉陷，可用于生态驿站

SPONGE CITY

景观设计尺度下的雨洪管理

STORMWATER MANAGEMENT IN LANDSCAPE DESIGN SCALE

225

226 宜居城市下的水敏性生态系统
Water Sensitive Ecological System in the Livable city

水作为人类生活和城市发展的基本要素之一，已经成为城市顶层设计的一个重要部分。在亚洲现代快速的城市化进程中，水在城市中是作为一个非亲和力的自然因素（传统水利概念），但现在这种观念正在改变。当着手建造或规划城市区域时，城市设计者面对的不是单一的环境美化和城市建设问题，而是城市的有机发展问题。因此，海绵城市不是一个新的概念，它由来已久，是与"城市化"并存的城市与环境问题。

人们规划水系统的传统方式往往是着眼于城市的功能、技术和经济因素，城市规划师根据需求制定原则，所有与水相关的基础设施都在此原则下进行设计。人们对于海绵城市设计的理解是城市水环境规划并没有僵硬、固定的原则和分层体系，而应该将城市规划、水环境、场地的社会文化功能及气候地形特征进行整合设计，把多种元素聚集在一起，达到可持续发展的目标和效益。海绵城市是实现基础设施与城市环境可持续发展的基石。在海绵城市的建设中，应该把城市中产生的水通过近自然的方式加以管理，让自然的水与自然环境形成有机的自循环。

景观规划设计需要适应全球气候和环境的变化，将景观规划师的使命与整个地球生态系统联系起来将是平行学科融合的必然。如今，海绵城市建设和城市发展结构的调整使景观规划师的职业明显不同于传统的角色定位。海绵城市设计需要运用多学科交叉配合的工作模式，打造不同尺度的高品质城市空间。

海绵城市的目标与营造更加安全、更宜居的城市生活相一致。它的大背景是节约和保护水资源，最大限度的实现环境的有机循环。海绵城市同时聚焦于水安全，其目标是为城市排水系统减少雨水径流、削减洪峰、提高水质。

海绵城市的一个重要转变，是将传统的地下基础设施变为可见的地上元素，目前，美国、德国、新加坡和中国的一些地方都已经实现了这种设计。传统的水管理设施中往往有一些外露的排水系统。景观设计师的任务就是美化它们，使其既具有城市基础设施的使用功能，又和谐的融入城市景观。

同时，海绵城市的建设存在两个难点和关键。第一个是教育层面，即怎样让从事规划设计的人员和政府人员更好的认识海绵城市的关键点、建设目标以及建造原因，这恐怕是现在很多地区尚未解决问题。第二个是技术层面，怎样在已建区域和新建区域中进行海绵城市的设计。这是一个非常有难度的问题，因为它直接涉及当地文化。每个项目都具有特殊性，设计也需要营造一种文化，且需要考虑它的过去与将来。海绵城市的建设需要多学科的共同参与，从规划、建造到最终完成，都不仅仅是设计师的任务。我们要考虑许多不同团体的声音，因为每一方都对项目有不同的认识和理解。将各方案集在一起并对他们进行教育，这样我们才能从同一起点出发，进而达到新的高度。每个人都可能对解决方案有不同见解，因此需要通过交流切磋达成共识，找到一个折中的方法，为特定区域设计出最好的方案。

如果城市水环境的规划应该契合城市文化和自然环境，满足现代人居环境要求，那么进行宜居城市的生态设计应该是所有专项设计的整体核心价值。一个可持续的宜居城市不仅需要形成具有雨洪管理能力的海绵城市，更需要把城市中的公共基础设施用多学科结合的方式整合到城市设计与建设管理中。如果海绵城市是解决城市发展中的水问题，那么宜居城市则是集水环境、生态景观及人文于一体的可持续、生态型现代城市发展结构。

这种新生代的城市发展结构，满足了城市雨洪管理、市民休闲娱乐、环境安全和生物多样性等多方面的城市建设要求，以极具战略性且可操作性强的方式长效保障城市的弹性适应能力和经济增长能力。具有海绵城市基础设施特色的互动型宜居城市，让我们的城市更美丽、功能更齐备，安全与舒适共存，形成可持续、交互式、公众性、环境友好型的社会结构。

安博·戴水道

Water, as one of the basic elements in human life and urban development, has become a crucial part of the urban high-end design. During the process of rapid urbanization in modern Asia, water is regarded as an un-amiable natural element (according to the traditional water concept), but now we can see that this concept is gradually changing. When begin to plan &build an urban area, the issues designers faced are not simply the city beautification and urban construction, but an organic development of the city. So, Sponges City isn't a new concept, but with a long time origin. It's a problem existed at the very beginning of urbanization.

Traditionally, the methods of water systems planning tend to focus on the city's functional, technical and economic factors, urban planners set out principles according to demands, of all water-related infrastructure should be designed following these principles. In the view of the Sponge City design, there're no rigid or unchangeable guidelines and layered systems in the urban hydraulic planning. It needs to integrates various elements, such as urban planning, hydraulic environment, social &cultural function, local climate and typographic features, etc., in order to achieve the objectives and benefits of sustainable development. Sponges City is a cornerstone to accomplish the sustainability of urban infrastructure and environmental. In the sponge cities, we should manage the urban-produced water in a biomimetic manner, to form a spontaneous organic circulation between natural water and environment.

Landscape planning & design needs to adapt to the global climate and the environment change. Linking the landscape planners' mission and entire planet's ecological system together is an inevitable trend in the relevant & parallel discipline integration. Now, adjustments in the Sponge City construction and urban developing structure, makes the landscape planners'occupation distinguished significantly from the traditional ones. Sponge City design requires a collaborative multi-disciplinary working mode to create high-quality urban space in different scales.

The goal of Sponge Cities corresponds with goal of creating safer and more livable city life. Its background is water resources conservation and protection, maximizing and smoothing the environmental organic circulation. Sponges City also focuses on water security. Its target is to reduce urban stormwater runoff volume, clip the flood peak, clear water quality in the drainage system.

Sponges City is a significant shift-point, it visualize the traditional underground infrastructure elements up on the land. United States, Germany, Singapore, and some parts of China, have already realized this kind of design nowadays. We can easily find some exposed drainage system in traditional water management facilities. Landscape architect's task is to attach aesthetic value on them, not only complete the necessary infrastructure construction, but also incorporated the hydraulic utilities into the urban landscape creation.

Meanwhile, there are two difficult &crucial points in Sponges City construction. The first is education,such as how to make the professionals and government officials who are occupied in planning & design work, be aware of the significance of the key points, building goal and construction reason of Sponge City. This is probably an unsolved problem in lot of areas yet. The second is technology and technique. such as how to implement Sponge City system in the newly build and already built region. This is a very tough question, because it's directly involved with local culture. Each project is unique, our design meant to create a culture, designers need to consider its past and future. Sponge City construction is a multidisciplinary participation. From the very beginning plan, construction to the final completion, they are not just the designers' efforts and tasks. Designers have to consider the voices and wishes from different groups and communities, because each party may have a different understanding and recognition towards the project. Designers firstly need to gather and educate all the parties so that we can set out from a unified starting point, and achieve a new perception height together. Everyone may have a different understanding of the solution plan. Therefore, the communication and negotiation are the premise to reach a consensus, in order to, put forward the most suitable & solvable project for a specific area in a compromise way.

If the urban water environment is to meet the needs of respecting the urban cultural & natural environment, satisfying modern living environment's requirements, conveying the comprehensive core value of every specialized plan in the ecological livable city design. Then a sustainable livable city should not only upgraded into the sponge city embodied with stormwater management capacity, but integrated the public infrastructures in urban design & construction management through a multidisciplinary approach. If the Sponge City is dedicated in solving reclaimed water during urban development, then the livable city is a modern ecological &sustainable city developing structure combined with hydraulic environment, ecological landscape and humanistic society.

This new generation of urban developing structure could meet the basic urban construction requirements, such as stormwater management, public recreation and entertainment, environmental security and biological diversity, etc. Besides, it could feasibly and strategically guarantee the cities' resilient adaptability and economic growth buoyancy in a long-lasting way. An interactive and livable city with the features of Sponge City infrastructures, make our city more beautiful, more multi-functional, more safety and more comfort, and form a sustainable, interactive, public and environment-friendly society structure.

RAMBOLL STUDIO DREISEITL

(Translated by Zhang Xu)

228
合肥融科城带状公园与水道设计

LINEAR PARK AND WATERWAY OF RAYCOM CITY, HEFEI

项目位置：中国安徽省合肥市
项目规模：50 ha
设计公司：SWA集团

Location: Hefei, Anhui Province, China
Size: 50 ha
Design Firm: SWA Group

合肥融科城是合肥市的一个新型经济开发综合片区,其中心区域是一个两侧设有零售底商步道的带状公园,这条舒适便捷的步行主轴线连接着该地区的8个街区、各类设施以及即将开通的地铁站。

区域性特征和汇水区域规划

合肥融科城所在之处地势平坦且缺少特色,原场地内多为成排的农田和零散的聚落。考虑到合肥的气候及全年的降雨分布情况（年平均降水量1 000 mm,每月降雨约5~12天）,SWA集团依据合肥标志性的河湖花园区域历史,探索项目的水文设计,旨在将该地区打造成契合当地降雨特色的场所。为了实现这一概念,SWA集团通过塑造地形来收集地表雨水径流,种植繁茂的树木造多样化的户外空间,为附近拥有100多栋高楼的街区营造一片人性化的生态绿洲。

SWA集团接受委托时,该地区街道的图纸虽已经完成绘制,但尚未筹建。经过初步研究,大部分拟建的带状公园、与之平行的街道和零售步道的排水管道都过于零散,且过度依赖管线,但仍有条件整合为单一的汇水区域。由于这一地区还尚未建设,所以景观设计师提出了用一系列雨水花园与水道来替代水管网的想法。SWA集团整体性地考虑户外空间的中央主轴线,调整街道和带状公园的竖向坡度,使商业街道、零售底商步道以及带状公园共同形成相互依存的汇水区网,既可减缓和净化雨洪,又可丰富人们的社区亲水体验。

采用绿色基础设施,而非灰色管网

合肥融科城带状公园的最终设计成果包括一条长780 m、深2 m的下沉式绿地水道,用来汇集和输送来自公园东部80%的地区、中央公共街道,以及两侧的零售底商步道之间的区域地表径流。部分邻近社区裙楼和层屋顶的雨水也可通过建筑排水管道直接流入水道系统。长780 m的水道中包含四个由低坝构成的中型池塘和一个位于东部低地的大型蓄滞水池,蓄滞水池中的小型排水管道在超出常水位时缓慢地将雨水排放到市政雨水管。在其上方1 m处,有一个更大型的紧急溢流结构,以应对大型暴雨。

由于下沉式绿地水道的高程比邻近街道低2 m,其收集雨水的能力与管网相同,因此能将其取而代之,如此一来,水道上方架设的与周边街道相平的桥面可供行人和车辆通行。

针对当地具体情况的设计

截至2015年5月,位于东部的带状公园下游部分和相邻两个社区中长约460 m的水道已经建设完成。其余部分预计最晚于2016年竣工。

与大多数位于大城市下游的水体一样,巢湖水质因受到城市建设带来的工地淤泥、城市面源污染物和农业径流的影响而不断恶化。为了净化雨水径流,合肥融科城运用植被对径流进行过滤,并沉降悬浮物质。值得一提的是,已建成的部分带状公园和水道已对该区域和在建的邻近区域中的建筑活动带来的大量施工淤泥进行了沉淀。当总长780 m的水道和侧翼的街区也建成时,带状公园将逐步实现水文生态平衡,带状公园和水道的功能将从沉淀施工淤泥转为沉淀城市径流中的颗粒物及减缓暴雨峰值。虽然这一地区并不与自然水道或城市河流直接相连,而是溢流向市政雨水管网,但其已促进了水质净化并减少了洪涝峰值流量。当社区全部建造完成之后,SWA集团计划选取由路缘流向水道的典型道路径流以及水道出水口的水质进行抽样检测,以监测水道的长期绩效。

为了减小该地区在由农业地区转变为城市区域的过程中的排放量,该设计赋予了水道了强大的蓄滞能力,并仔细平衡了该地区铺装区域与种植区域的比例。鉴于场内主要为黏性土,透气性和渗透性均较差,该设计未将提高水道的下渗能力作为主要设计目标。

随着该区域设计和施工的快速进行,SWA集团作为规划设计者应该及时反应,为带状公园和水道提供令人信服的设计和技术细节支持。在接下来的后续项目中,SWA集团将与环境顾问和工程师深入协作,更好地了解并监测项目中的景观绩效,进一步调整水道中的植被配植比例和植物种类性能,并利用各方反馈讯息对池塘布局等方面进行持续优化设计。

Raycom City, a new mixed-use area in Hefei, has a public linear park flanked by retail promenades as its heart. The linear park and retail promenades provide a comfortable, convenient walking spine connecting the area's eight neighborhoods, various facilities and amenities, and an upcoming metro station.

Regional Identity and Watershed Planning

The site is generally flat and featureless recently with row crop agriculture and small clusters of homes. Given Hefei's climate and rainfall distributed throughout the year (1,000 mm average annual precipitation on 5 ~ 12 rain days per month), SWA Group explored concepts that drew on Hefei's iconic river parks to provide a distinct sense of place for the site. To realize the concept, SWA sculpted the site topography to capture on-site rain water flows and create a variety of outdoor spaces by fostering thriving canopy trees, providing a leafy human-scaled heart shared by the people living over 100 towers in the neighborhood.

When Raycom commissioned SWA Group, the plan of the streets in the area had been drawn but not yet realized. At first glance, most of the planned linear park, parallel streets and retail promenades appeared to potentially form a single watershed. Because the site was undeveloped, landscape architects were able to propose a series of rain gardens and a waterway in lieu of storm pipes. SWA Group then started to think about the central outdoor spine holistically and adjust grades for the streets and the linear park to form a single self-draining watershed storefront-tostorefront that could both slow and cleanse stormwater as well as enrich the people experience of the neighborhood.

Applying Green Infrastructure, Rather than Grey Pipes

The resulting design for Hefei Raycom City's linear park includes a 780 m long, 2 m deep waterway, which conveys all rainwater which falls on the eastern 80% of the park parcels and central public streets, as well as the flanking retail promenades to the storefront setbacks of the adjacent private parcels. Drainage from a portion of the podiums and tower roofs of adjacent neighborhoods will also be directed into the waterway via building drain pipes. The 780 m long waterway includes 4 intermediate ponds formed by weirs, as well as a large receiving basin at the eastern low land where a small outflow pipe slowly releases drainage into the municipal storm pipe when the waterexceeds the ordinary water level. A larger overflow structure, 1 m above the outflow pipe,responses to the heavy storm events.

By setting the waterway flow line 2 m below the adjacent street and walkway paving, the open waterway is able to serve the same function as the in-street storm pipe it replaces, and to accommodate at-grade bridges for walking and vehicle access.

Designs for Local Conditions

By May 2015, the east, lower half of the linear park and adjacent two neighborhoods have been built including a 460 m long part of the waterway. The west half of the linear park and remaining upstream portion of the waterway is anticipated to be completed by 2016.

If Chao Lake is like most water bodies downstream of large cities, its water quality is impacted by city construction silt, urban pollutants, and agri-

GREEN STREET · STREETSCAPE DRAINAGE · LINEAR PARK & WATERWAY

cultural runoff, so Hefei Raycom City's strategies to enhance stormwater quality by filtering flows through vegetation and settling out particulates can only help. Anecdotally, the initial linear park and waterway have settled-out significant construction silt from ongoing construction activity on this and adjacent sites. When the waterway finishes construction and the environment of the linear park reaches balance, and the flanking neighborhoods are built-out, the linear park and waterway will shift from settling out construction silt to settling out particulates from urban runoff and slowing rainfall peaks. Since the site does not touch a natural waterway or city canal, the waterway overflows into municipal storm pipes, but with cleaner water and reduced peak flows. Once the neighborhood is fully built, we plan to sample and test typical street runoff at a curb cut inflow point to the waterway as well as water at the waterway outflow, to measure the waterway's performance in the long-term condition.

To slow drainage rates as the site changes from agriculture to an urban area, SWA Group sculpted significant capacity into the waterway, as well as carefully balanced paving and planting within the site's heart. Given the site's clay soils, permeability and percolation seemed less of an option and were not a major pursued here.

With the project's design and construction proceeding at a fast pace, SWA Group had to react quickly to create a compelling design and supporting technical details for the linear park and waterway. In the next broadly similar project, SWA Group will further explore and collaborate with environmental consultants and engineers to even better understand and monitor the effectiveness, further fine-tune the proportion of vegetation and vegetation species in the waterway, and further test waterway including pond configurations to continue improving their performance.

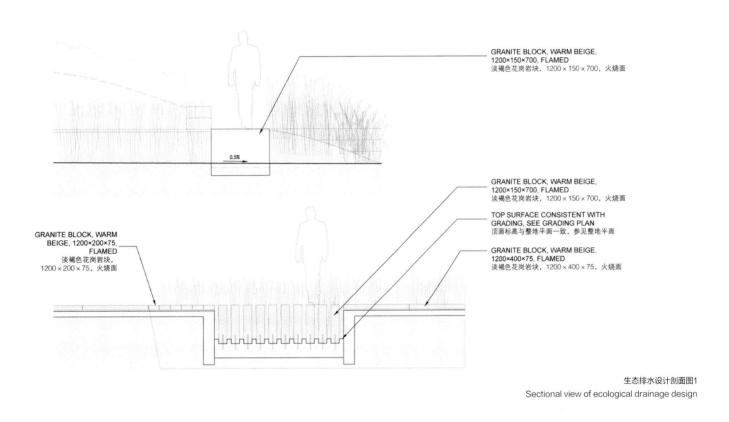

生态排水设计剖面图1
Sectional view of ecological drainage design

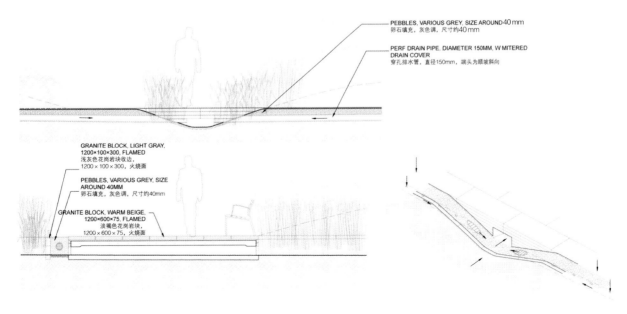

生态排水设计剖面图2
Sectional view of ecological drainage design

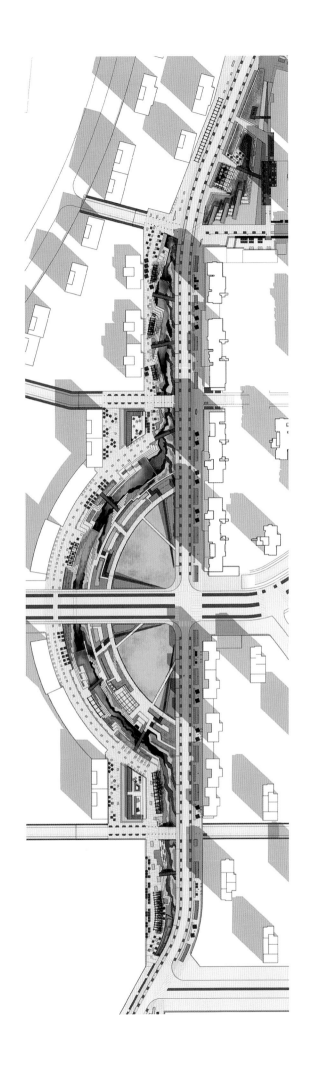
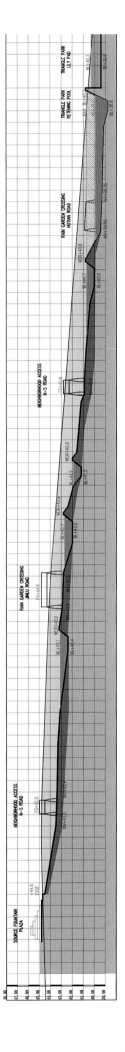

236

格思里城市绿地

GUTHRIE URBAN GREEN GARDEN

项目位置：美国俄克拉荷马州塔尔萨市

项目规模：1 ha

设计公司：SWA 集团

获奖信息：活动中心杰出设计奖；获得提名，ULI城市开放空间奖；

棕地更新奖；WAN城市设计奖；ASLA南加州区域荣誉奖；

亨利·贝尔曼（Henry Bellmon）可持续奖。

Location: Tulsa, Oklahoma, USA

Size: 1 ha

Design Firm: SWA Group

Awards: Center for Active Design Award of Excellence;
Finalist, ULI Urban Open Space Award; Brownfield Renewal Awards;
WAN awards Urban Design; ASLA Northern California Chapter Honor Award;
Henry Bellmon Sustainability Award.

公园所在地曾经是装卸码头,这里全年提供治愈和健身课程,同时也被用作艺术表演、电影放映和农贸集市的场地。现在由乔治·凯尔撒(George Kaiser)家族基金掌管并打理。

该1.05 ha的高效能公园拥有地源热泵系统、生物沟、可亮化整个社区街景的LED照明和一片多功能草坪。1022 m² 场馆中的光电纵列是可再生性能源的稳定供应源。生物沟可处理雨水径流、过滤污染物质并且能够补给地下水源。在过去的三年里,公共私人基金组织联合投资总额高达1.135亿美元,保证了公园中地理交换系统的维修养护资金。

In the past, the park was a loading dock: it provides year round free programs of health fitness classes and art, film and farmer market events. Now, it is and maintained by the George Kaiser Family Foundation (GKFF).

The 2.6-acre high-performance park includes a ground source heat pump system, bioswales, district wide streetscape with LED lighting, and a multipurpose lawn. Photovoltaic arrays on the 1,022-square-meter pavilion supply a steady source of renewable energy. Bioswales treat storm water runoff and filter contaminants while recharging groundwater sources .Over the last three years, a total of $113.5 million has been invested through a combination of public-private funding and grants to support the park's geo-exchange system.

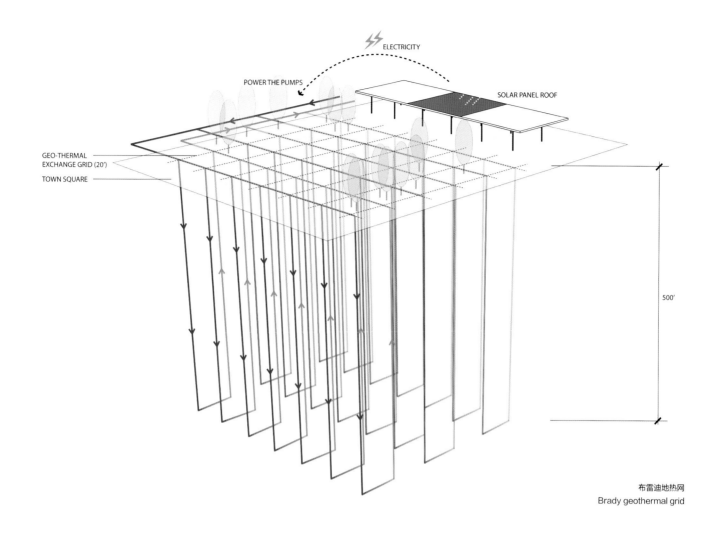

布雷迪地热网
Brady geothermal grid

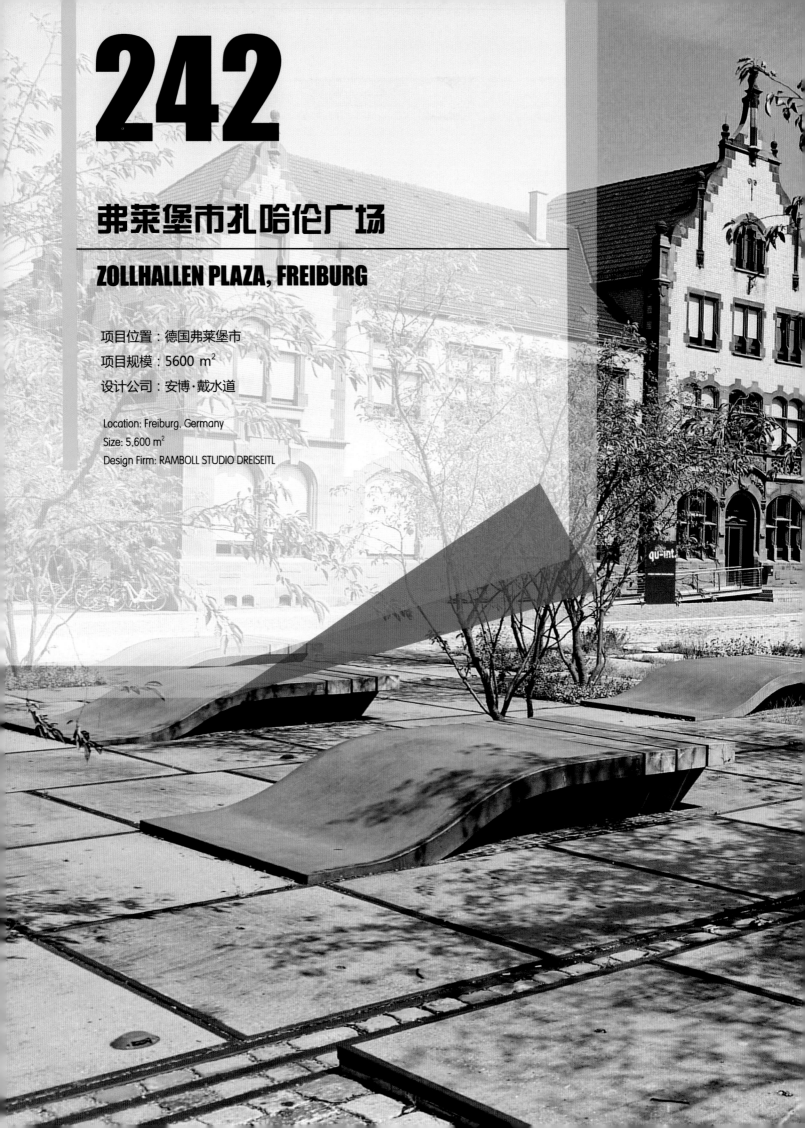

242

弗莱堡市扎哈伦广场

ZOLLHALLEN PLAZA, FREIBURG

项目位置：德国弗莱堡市
项目规模：5600 m²
设计公司：安博·戴水道

Location: Freiburg, Germany
Size: 5,600 m²
Design Firm: RAMBOLL STUDIO DREISEITL

扎哈伦广场的建设为：2009年历史悠久的海关大厦的修复相得益彰。该广场完全摆脱了污水处理系统，成为水敏性城市设计的一个典型案例。美丽的种植池提供了渗透点，拥有创新式内置过滤基质的地下砂石沟渠减轻了污水处理系统的水压负载。在缩进的广场区域中创建了一个地表防洪区。雨水没有汇入地下污水处理系统，而是补给地下水位。该广场建造于曾经的铁路院落之内。富有年代感的多功能座椅唤起了人们对于铁路轨道枕木的记忆，而旧的铁路轨道则嵌入地面铺装之中。一片明快的樱花树林提供了充足的树荫，渗透式种植池之中育有多年生植物和观赏草类，形成了鲜艳、柔美的花草景观。100%的硬质景观材料均来自旧的铁路院落场地回收利用的优质材料。因此，在资源管理和建造材料来源方面，这个现代化的广场与历史悠久的海关大厦相互映衬。

Zollhallen Plaza is new counterpart for the historic customs hall which was restored in 2009. The plaza is a fine example of water sensitive urban design, as it is disconnected from the sewer system. Beautiful planters provide infiltration points, and subsurface gravel trenches with innovative in-built filter medium reduce the hydraulic overload on the sewer system. Indented plaza areas create a surface flood zone. No rain water is fed to the sewer system, instead the groundwater table is recharged. The design plays with the historic past of the site which was a railyard. Timeless and multifunctional benches recalled break noses of railtracks, and old railtracks are inlaid into the paving. A bright grove of cherry trees provide the perfect amount of shade, while the infiltration planters with perennials and ornamental grasses give an attractive softness. 100% of the hardscape materials are high-quality demolition materials recycled from the old railyard. This makes sense not just from a resource management point of view, but harmonises the new clean modern design with the historic architecture of the customs hall.

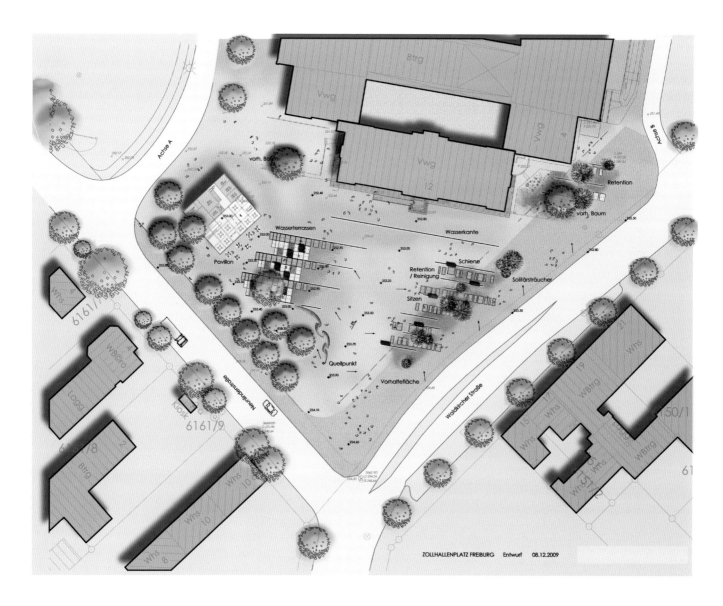

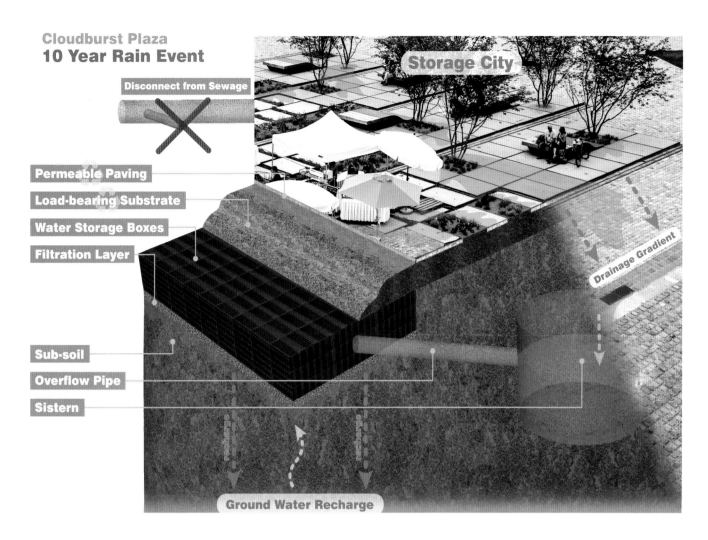

Cloudburst Plaza
10 Year Rain Event

- Disconnect from Sewage
- Permeable Paving
- Load-bearing Substrate
- Water Storage Boxes
- Filtration Layer
- Sub-soil
- Overflow Pipe
- Sistern

Storage City

Drainage Gradient

Ground Water Recharge

248

西雅图文化广场

SEATTLE CULTURE SQUARE

项目位置：美国华盛顿州西雅图市
项目规模：640 m²
设计公司：安博·戴水道
翻　　译：张旭

Location: Seattle, Washington, USA
Size: 640 m²
Design Firm: RAMBOLL STUDIO DREISEITL
Translated by Zhang Xu

西雅图文化广场是西雅图文化中心总平面的最后一块，这里将被建设成综合大厦，与毗邻的广场合为一体，提供各种舒适的零售,办公和居住空间。大厦内部的空间及外部开放的广场,可容纳一系列活动,包括个人娱乐和团体集合。

开放空间中的各部分被巧妙地整合在一起,形成了"大姿态"的城市景观。

塔状屋顶与原本用作支撑物件的硬质铺装上的雨水径流成为可持续雨洪管理系统的组成部分。

Seattle Culture Square is the final piece of the Seattle Culture Center master plan. When finished, the one city block, with its mix-used tower and adjoining public plaza, will offer a variety of experiences through its retail, office and residential amenities, within the tower as well as the flexible open space of the plaza, designed to accomodate a range of activities,- from large events to individual enjoyment of the space.

The open space layout is planned as an urban landscape sculpture. Its parts are integrated to create the composition of the open space and together seen as the "big gesture".

Stormwater from the tower rooftop and surrounding hardscape used to help supply these elements will be a part of the sustainable stormwater management system.

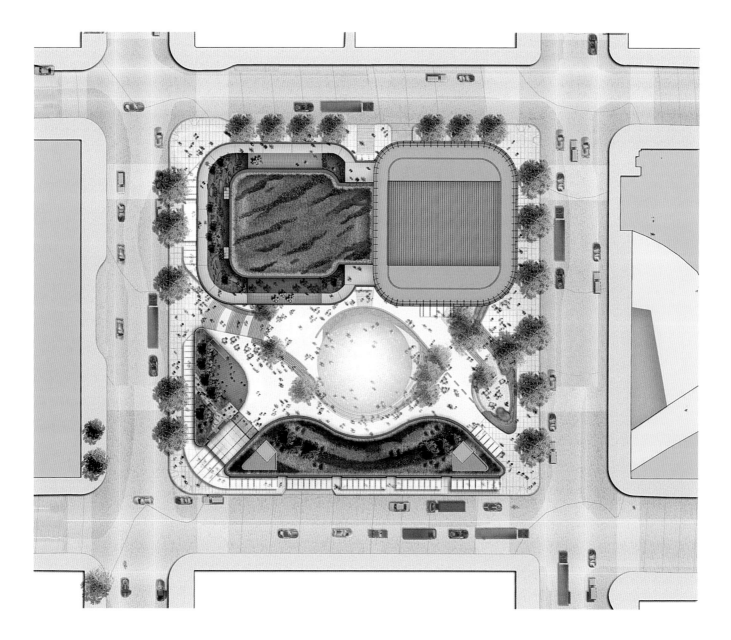

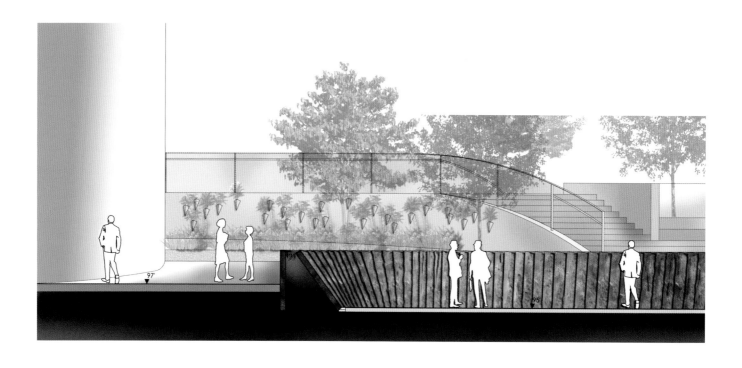

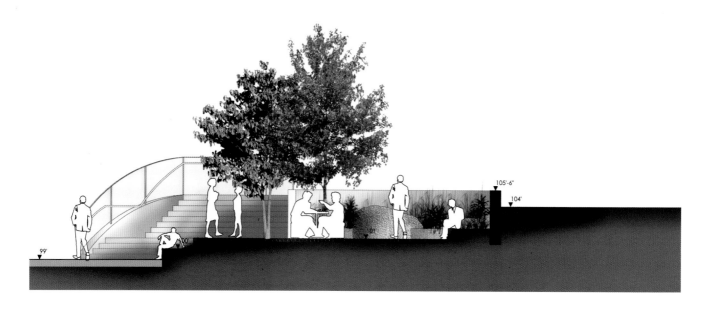

254

埃纳河的日光：奥斯陆海伦拉卡中心公园

DAYLIGHTING OF ALNA RIVER CENTRAL PARK: HOLALLOKA, OSLO

项目位置：挪威奥斯陆市
项目规模：2 ha
设计公司：安博·戴水道

Location: Oslo, Norway
Size: 2 ha
Design Firm: RAMBOLL STUDIO DREISEITL

该项目结合了雨水管理与非正式休闲城市公园的功能，是大型的埃纳河恢复项目的一个试点工程。埃纳河从老旧混凝土管道之中解脱出来，被修复成拥有软质植被河岸和池塘区域的自然形式。这样的"再自然化"形成了独具特色的绿色公园，不仅为动物们提供了良好的栖息地，同时也可有效地发挥河流的管理功能。

高质量和艺术化的设计将修复河流的自然本色整合到城市整体结构中。新路径连接着河流与当地道路交通，提供了新的娱乐场地。河流水质良好，人们甚至可以游泳！拥有岸滩的游泳池成为公园的中心焦点。木制甲板是一个舒适的日光浴场地，人们在此可以观景；同时，木质甲板还可作为踏步石以及拥有大大小小排水口的水瀑元素。水管理的工程技艺以幽默和充满才华的方式被展示出来，营造了轻松愉悦的氛围中。相邻工业建筑的雨水管理与河流系统有效结合。从建筑屋顶上收集雨水，雨水通过地表的细节处理流入一片植物生态净化群落，并通过多层土壤基质的过滤得到净化，放缓速度，缓慢流淌。

The project is combined with stormwater management and an urban park with informal recreation areas, as a pilot for the large-scale restoration of Alna River. The Alna is released from its old concrete pipe and restored to its natural form of soft planted banks and pond areas. This re-naturalization not only provides habitat creation and a distinctive green character to an urban park, but also ensures important river management functions.

High-quality and artistic design integrates the restored natural qualities of the river into the urban fabric. New pathways link the river into local circulation and provide new recreation possibilities. The quality of the water is good enough to swim in! A bathing pond with a beach is the central focus of the park. A wooden deck is a comfortable sunning area with views to cascade elements which double as stepping stones and various detention outlets. The technical aspects of water management are integrated with humour and flair into the attractive and relaxed atmosphere of the park. The stormwater management of the adjacent industrial buildings is combined into the river system. Rainwater is collected from the roofs and conveyed via surface details into a planted cleansing biotope area. Here, rainwater is cleansed through a multi-layered substrate and held back for slow release.

Cleansing Biotope

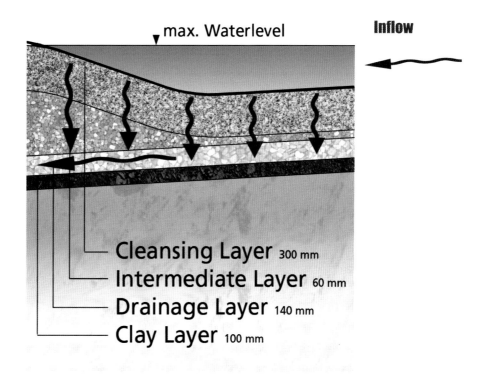

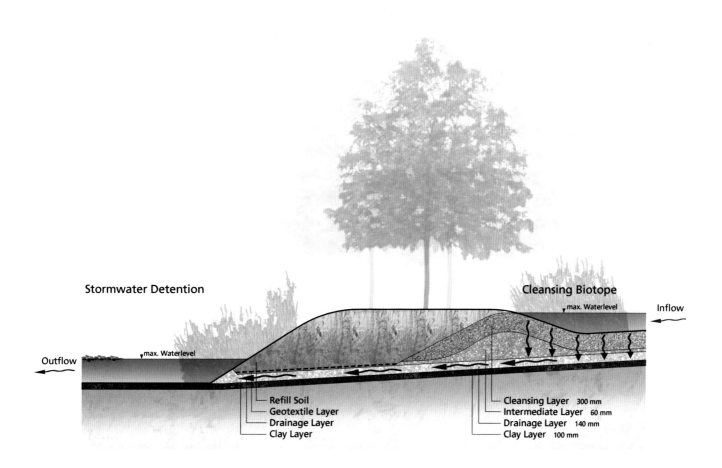

260
巨人网络集团总部景观规划设计

LANDSCAPE PLANNING AND DESIGN OF GIANT INTERACTIVE GROUP

项目位置：中国上海市
项目规模：18.2 ha
设计公司：SWA 集团
翻　　译：张旭

Location: Shanghai, China
Size : 18.2 ha
Design Firm: SWA Group
Translated by Zhang xu

巨人网络集团总部位于中国上海，其18.2 ha的园区和绿色屋顶，被构想为生态公园和活态实验室。该项目依据自然系统与开放空间的规划进行建设，景观与建筑在其中衔接得天衣无缝。一半以上的园区空间拥有合作办公功能，在另一半的园区空间内，生活方式的经营是关注重点，其中包括宾馆、会所以及餐饮和休闲空间。

SWA集团试图用景观组织场地空间的框架，其中包含一系列聚焦生态可持续性的亲水体验。该基地周边被农业灌溉河道环绕，这里原本是香樟树（番樟属，樟树）和桂树（木樨属，桂花）的育殖园区。现有的许多树木经过补救养护后，被移栽于园区内的不同位置。城市道路（中凯路）横穿这片土地，将它一分为二。基地规划中以水作为连接组织不同景观项目的元素。繁杂精巧的水文系统覆盖了现有的灌溉河道、新建的滞水坑塘和季节性湿地，它们为野生动物提供了多样化的栖息环境，同时也可被用作科研试点。园区中，水生栖息地与湿地栖息地将不同的建筑要素编织其中。

巨人网路集团周边水道密布，跨越中凯路南的总部大楼景观壮美。15,222 m²的绿色屋顶耗资昂贵，覆盖了整栋建筑结构，过渡并软化了建筑与景观边缘之间的生硬衔接。屋顶天台的地势在公司的招待宾馆处达到最高高程，其中的私家休憩套间悬浮在湿地池塘之上。其他活动与额外项目位于绿色屋顶之下，例如多功能运动场地、健身区域。

绿色屋顶被设计为低成本养护"草地"，长时间内不用灌溉与施肥。不似常规意义上的屋顶花园，草皮不会出现褶皱、突起、下陷的情况。起伏绵延的屋顶结构衔接入地表平面、延伸进邻近野生池塘，同样也连接者步行广场。最极端的斜面倾斜程度高达53°，使得植被栽植工程充满挑战。整个创新加固型混凝土楔铸系统，通过钢筋剪角与石笼形成共轭，与每一个斜置的屋顶表面结构平行。该系统形同巨型独立细胞，使土壤保持在原位，以降低重力导致的下沉与腐蚀。尺度巨大的屋顶正如一个蓄热体，限制了热量的吸收也减少了制冷的支出。起伏折叠使屋顶不同区域的几何特征与朝向产生差异，因为背脊、沟壑、阳光、阴影等因素衍生了迥然不同的微气候环境。经过一年的植被甄选与栽植配比的实验，基于植物的源产地、抗倒伏性、喜光性、耐荫性、耐旱性、耐水性和散花物候期选择了11种植物（其中有7种源产于中国）。为防物种竞争，每种植被的成活率也被充分考量。仅仅一个季度的生长之后，屋顶景观空间成为了蝴蝶的天堂，尤其是对于小型菜粉蝶（粉蝶属，菜粉蝶）而言。

沼泽和岛屿系统中的滨水树木、水生植被和休憩湿地，构成了平衡工作与生活的舒适的缓冲地段。最初的项目规划中包含临近总部大楼员工宿舍，两处建筑之间由运河分隔。与景观交相呼应的广场为员工提供了户外和休闲空间，中央通勤主道和连续露天步道通往湖边，并为人们提供了集聚机会。基址上景观、建筑和环境的无缝衔接是整个设计的最终成果。

与园地相融的广场是休闲放松区域。
Plazas are incorporated on the campus to provide areas of respite and relaxation.

Planting

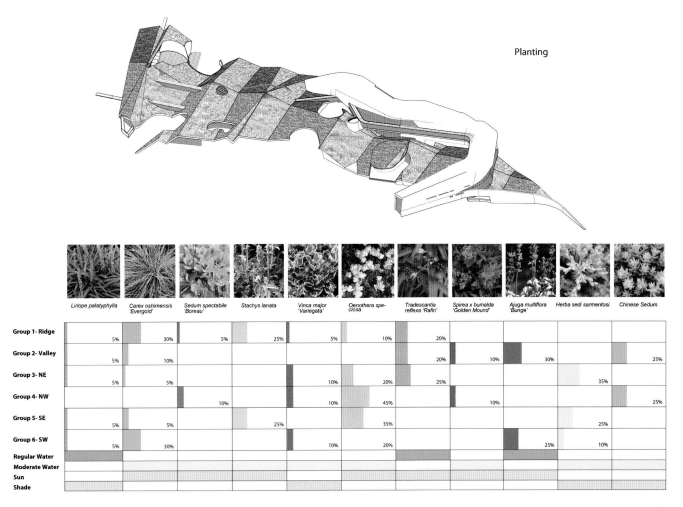

	Liriope palatyphylla	Carex oshimensis 'Evergold'	Sedum spectabile 'Boreau'	Stachys lanata	Vinca major 'Variegata'	Oenothera speciosa	Tradescantia reflexa 'Rafin'	Spirea x bumalda 'Golden Mound'	Ajuga multiflora 'Bunge'	Herba sedi sarmentosi	Chinese Sedum
Group 1- Ridge	5%	30%	5%	25%	5%	10%	20%				
Group 2- Valley	5%	10%					20%	10%	30%		25%
Group 3- NE	5%	5%		10%		20%	25%			35%	
Group 4- NW			10%		10%	45%		10%			25%
Group 5- SE	5%	5%		25%		35%				25%	
Group 6- SW	5%	30%			10%	20%			25%	10%	
Regular Water											
Moderate Water											
Sun											
Shade											

种植图解：混合种植的本土物种与适生物种覆盖在总部建筑表面，根据阳光朝向与水分需求分组种植。

Planting diagram: A mix of native and adaptive species cover the headquarters building and are grouped based on sun orientation and water requirements.

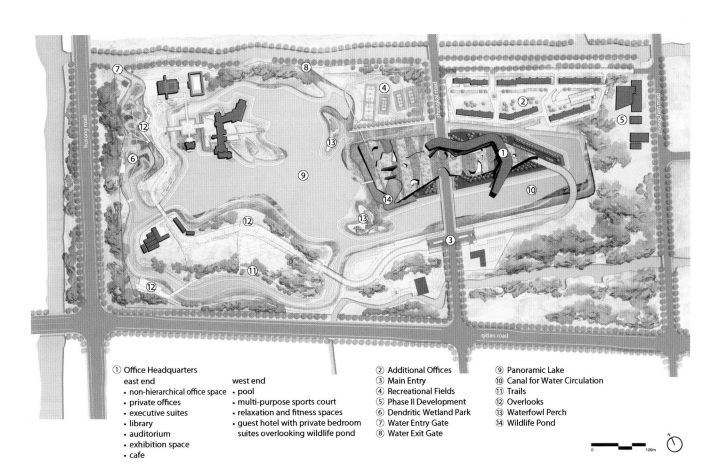

① Office Headquarters
east end
- non-hierarchical office space
- private offices
- executive suites
- library
- auditorium
- exhibition space
- cafe

west end
- pool
- multi-purpose sports court
- relaxation and fitness spaces
- guest hotel with private bedroom suites overlooking wildlife pond

② Additional Offices
③ Main Entry
④ Recreational Fields
⑤ Phase II Development
⑥ Dendritic Wetland Park
⑦ Water Entry Gate
⑧ Water Exit Gate

⑨ Panoramic Lake
⑩ Canal for Water Circulation
⑪ Trails
⑫ Overlooks
⑬ Waterfowl Perch
⑭ Wildlife Pond

The 45 acre campus and green roof for Giant Interactive Group in Shanghai, China was conceived as an ecological park and living laboratory. Structured around a plan of natural systems and open space, both the landscape and architecture programs seamlessly integrate. Half of the campus contains corporate and office uses, while the other half is focused on lifestyle, and includes a hotel, clubhouse, along with dining and recreational spaces.

SWA sought to use the landscape as an organizing framework for the master plan with a variety of water experiences focused on ecological sustainability. The site was formerly a tree nursery for mature Camphors (Cinnamomum camphora) and Sweet Olives (Osmanthus fragrans) surrounded by agricultural canals. Many of the existing trees were salvaged and relocated across the campus. A city road (Zhongkai Road) bisects the land, splitting it into two sections. Planning of the site employed water as a means to connect and organize various program elements. An intricate hydrological system consisting of existing irrigation canals, new water retention basins, islands, and seasonal wetlands, created a diverse habitat for wildlife as well as scenic points. Water and wetland habitats weaved together the various architectural elements on the campus.

Surrounded by waterways, the headquarter building for Giant Interactive Group spans across Zhongkai Road South to create a powerful impression. An expansive green roof of (15,222 square meters) envelopes the structure, blurring the edges where landscape and building meet. The rooftop landform culminates in a company guest hotel where private bedroom suites project over the wetland pond. Additional program is beneath the green roof in the form of a multi-purpose sports court, and fitness areas.

The green roof was designed to be a low maintenance "meadow" that requires little watering and naturalizes over time. Unlike a typical green roof, the surfaces fold, soar and dip. The undulating roof structure touches the ground plane, dipping into the adjacent wildlife pond as well as coming into contact with the pedestrian plaza. The extreme slope conditions vary up to 53 degrees and posed significant challenges for vegetation. An innovative system of reinforced concrete cleats, spanned by steel angles and gabions, are laid parallel to each sloping surface. (Image 001) The system functions as large self-contained cells holding the soil in place and thus minimizing slumping and erosion due to gravity. The roof's extensive size acts as a thermal mass that limits heat gain and reduces cooling expenditures. Because of the roof's folding geometry and orientation, distinct micro-climates occur in response to ridge, valley, sun and shade conditions. After a year-long of testing plants and mix percentages, eleven species were chosen based on the plant's native origin (seven species are native to China), plant rigor, ability to tolerate sun and shade, plant hardiness to dry soil and standing water, and seasonal flowering regimes. The growth rate between plant species is also taken into consideration to prevent species competition. After only one season's growth, the roofscape has become a haven for butterflies, particularly the Small Cabbage White butterfly (Pieris rapae).

Riparian trees and plants, and a wetland sanctuary of networked marshes and islands make a comfortable buffer balancing work and lifestyle. The original program also includes employee housing set adjacent to the headquarters building and separated by a canal. Several plazas carved from the landscape provide outdoor and recreational spaces for employees, while a central circulation spine and continuous outdoor walkway provide access to the lake and opportunities for gathering. The overall design outcome is the seamless connection of landscape, architecture and environment to the site.

策略性地放置玻璃幕墙可让人们领略绿色屋顶的景致。
Glass curtain walls are strategically placed to give views of the green roof.

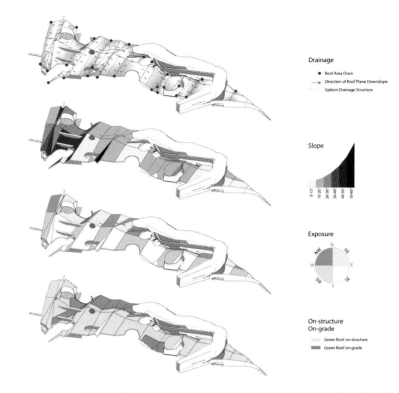

景观系统图解：排水、坡度、暴露程度和土壤厚度的不同呈现出巨型绿色屋顶整体风貌的不同。
Landscape systems diagrams: Variation in drainage, slope, exposure and soil depth contribute to the overall performance of the Giant green roof.

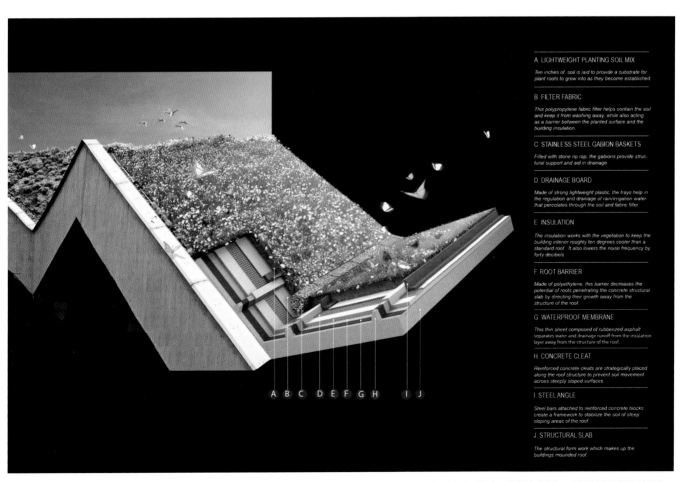

屋顶系统：绿色屋顶结构细部。建设排水、强化结构和可持续生长的综合系统。该系统在保证植被增长的前提下可控制腐蚀并且保护房屋结构。
Roof system Detail of green roof construction. An extensive system for drainage, structure, and sustainable growth was developed, supporting growing plants while controlling erosion and protecting the roof structure.

植被覆盖屋顶结构，丰富的花期吸引了大量菜粉蝶。
The roof structure is carpeted with prairie-like plants that have variable blooming seasons.

隐蔽的泻湖植被茂盛、草木葱茏并且有很多动物生活在其中。游人可以在建筑内部欣赏这个生态系统的景观。
An enclosed lagoon has diverse water grasses, plants and animals. Views of this ecosystem can be seen from visitors inside the building.

运河环绕在总部大楼与人行步道周边,人们可以在此休闲娱乐。
A canal surrounds the headquarters and pedestrian walkways, where recreational opportunities can be found.

建筑的玻璃幕墙模糊了"内部"与"外部"的界限,游人可透过其与自然进行静态互动。
Blending the lines of what is "inside" and "outside", the building's curtain wall of windows allow visitors to passively interact with nature.

在建筑北部，员工可用露天广场和行人步道。
On the north side of the building, employees can use open plazas and pedestrian walkways.

种植图解：混合种植的本土物种与适生物种覆盖在总部建筑表面，根据阳光朝向与水分需求分组种植。
Planting diagram : A mix of native and adaptive species cover the headquarters building and are grouped based on sun orientation and water requirements.

在建筑北部,员工可用露天广场和行人步道。
On the north side of the building, employees can use open plazas and pedestrian walkways.

建筑的悬臂结构前景展示了其与下沉式绿色屋顶的关系。
The front view of the cantilever structure from the building shows its relationship to the dipping green roof.

270
伊斯坦布尔格基奥马帕萨区城市规划和雨洪管理
URBAN PLANNING AND STORMWATER MANAGEMENT OF GAIIOSMANPASA, ISTANBUL

项目类型：市政景观规划，城市雨洪管理
项目位置：土耳其伊斯坦布尔市
项目规模：1100 ha
规划设计：福斯特
景观设计：安博·戴水道
翻　　译：张旭

Project Type: Municipal Landscape Planning, Urban Storm Water Management
Location: Istanbul, Turkey
Size: 1,100 ha
Mater Plan: Foster
Landscape Design: RAMBOLL STUDIO DREISEITL
Translated by Zhang Xu

城市河谷

格基奥马帕萨区最突出的特点是一条蜿蜒曲折的河谷把格基奥马帕萨与土耳其最重要的港口金角湾联系在一起。该河谷的特殊之处一方面是它的地形与众不同,另一方面是它具有很大的历史意义,因为它属于伊斯坦布尔供水系统的一部分。该系统历史悠久,由16世纪意大利文艺复兴时期最伟大的建筑师和工程师之一米玛尔·科卡·思南(Mimar Koca Sinan)设计。

结合经济发展和自然环境两个重要的城市设计要点,设计师把单一的城市河谷改造成了城市绿色走廊,为这个城市提升了娱乐和生态服务的潜力。这使城市河谷不再是一个纯线性结构的公园,且是一个集各种文化娱乐、水文管理和城市职能于一体的生态高效可持续的现代城市结构。当然,城市河谷的生态开发也将大大提升沿线建筑和房地产业的品质,为其带来更大的开发潜质和价值。

在总体规划中,城市河谷被看作是整个城市的蓝绿基础设施,设计师通过有效的城市水文和景观设计使它可减少城市洪水风险,改善水质,并对雨水进行回收和储存,以便在旱季加以利用。海绵城市系统设计不仅使城市具有调配水资源的能力并提升了水量,让生活在其中的城市居民享受良好的城市生活环境,而且改善了城市结构,为城市的经济发展带来了更多的活力。

城市林荫道连接了3个城市的三个主要区域:北部的新商业中心,南面的城市中心,南北之间的城市副中心。平行的城市河谷和城市林荫道为城市发展提供了更多的机会和有利因素。这两者的结合创建了经济发展和自然环境协同作用的良性城市发展结构。

邻里尺度

城市河谷中的雨洪管理措施主要分布于城市街道、餐厅、商店、轻工业企业等城市二级结构之中。二级街道是实现对水敏感的城市设计(被称为"可持续城市排水系统SUDS"或"水敏性城市设计WSUD")的理想区域,这将有利于降低地方和城市区域的污水管网水力负荷。绿色街道和环境友好性社区营造了有吸引力的街景,并显著地减缓了气候变化对城市环境带来的影响,创造了高性能的社会经济成本效益。

海绵城市设计/雨水管理系统

水资源的枯竭是城市作为一个整体所面临的最大挑战之一。历史上如此,随着城市化和水的需求增加,未来更是如此。该区总体规划的目标是成为城市水环境可持续发展体系的典范。

伊斯坦布尔的年降雨量超过伦敦(每年大约630~800 mm),当地温度和降雨量的区域分布不均是导致该地区5-9月干旱的主要原因。该方案力求储存丰水时期的雨水并在干旱时期用于城市用水和灌溉。

雨水管理系统(即海绵城市设计)推动了格基奥马帕萨区的生态可持续性发展,不仅是管理自己的水资源和水质,还扩展到储水用水,造福于城市。

宜居城市

城市景观设计和海绵城市设计的愿景是为城市发展提供更多的服务,使城市成为宜居城市。它综合更多生态的环境设计技术,更多的生物多样性,更自然的水景观,更好的水质以及更多的娱乐休闲区域。它使生活在密集城市环境中的人们拥有卓越的生活品质,也使该地区在面对气候变化和自然资源的压力时更有弹性。

成功的宜居城市建设需要投资与收益的平衡。宜居的环境显著地提高了当地地产价值,吸引高端人才与优秀企业,使当地经济健康持续发展。这就是人们所倡导的"建设高品质生活"的必要前提。

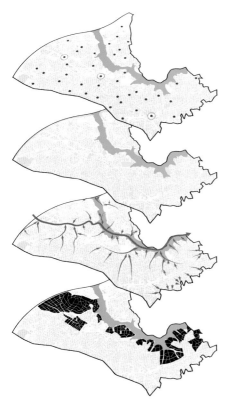

城市网络结构
Urban network structure

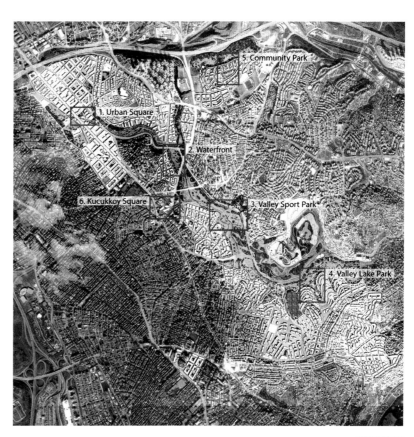

开发后平面图
Site plan after development

City River & Valley

The most prominent feature of Lattice Omar Pasa district Gaziosmanpasa is its winding valley which link lattice Omar Pasa and the Golden Horn Bay (the most important port of Turkey) together. The particularity of the valley is characterized both in the distinctive land typology and the historic significance, because it is a part of water supply system in Istanbu . This system boasts with its long history and brilliant cultural notion. It had completed in the 16th century by Mimar Koca Sinan who was the one of the most eminent Italian architects and engineers in the Renaissance period.

Combined the two vital urban design elements, namely the economic development and the natural environment, the designers Dreiseitl transformed the monotonous urban valley into an urban green corridor, improved and developed the potential for the function of recreation and ecological service. This makes the city an eco-efficient sustainable and feasible contemporary city structure combined with cultural entertainment, hydraulic management and multiple city functions, instead of just a linear-framed park. Of course, the ecological development along the urban river valley will greatly promote the quality of construction industry and real estate, in turn, bring them with bigger potential and better value.

The urban river valley is been perceived as a blue-green infrastructure throughout the city in the master plan. Through effective urban hydrology and landscape design, the designer from Atelier Dreiseitl makes it possible to reduce the risk of the flood in the inner-city and improve the quality of the water, recollect and store the rainwater with the purpose of reuse it in the dry season during summer. The Sponge city design in lattice Omagh Pasadena not only enable the city to embody the ability of regulating and allocating water resources, but also expand the water volume and quantity. All these can make the residents feel good about the urban living environment, and upgrade the city structure to stimulate the city's vitality in the economic growth.

The three main areas in the city are linked by the urban boulevard, namely, the new commercial center in north, the city center in south, the sub-civic center between north and south. The Paralleled valleys & urban boulevard provide the city with more opportunities and benefits for urban development. This combination of the valley & boulevard creates a healthy urban development structure in which economic development and the natural environment could behave in a synergized manner.

Neighborhood Scale

Stormwater management measures in the urban valley are mainly installed on the secondary city structure, such as the urban streets, restaurants, shops, businesses, light industries, etc. These sub-streets are the ideal sites to practice the water sensitive urban design (also called as Sustainable Urban Drainage Systems SUDS or Water Sensitive Urban Design WSUD), which will devoted in alleviating the stress in hydraulic loading of sewage pipe network in the local sits and urban areas. The green streets and environmental-friendly communities create an attractive and appealing streetscape, and significantly reduce the impact of climate change on the urban environment, bestow the city with a socio-economic cost-effectiveness of high-performance.

Sponge Urban Design / Stormwater Management System

Depletion of water resources is one of the biggest challenges a comprehensive city faced. The experiences in history had already verified this fact, and the situation in the future won't change in the terms of the undergoing urbanization and the increasing water demand. So the target of the master plan is to set a model for the urban water environment sustainable development system.

Istanbul annual rainfall overtops the one in London (fluctuate between 630 to 800 mm annually), the uneven temperature

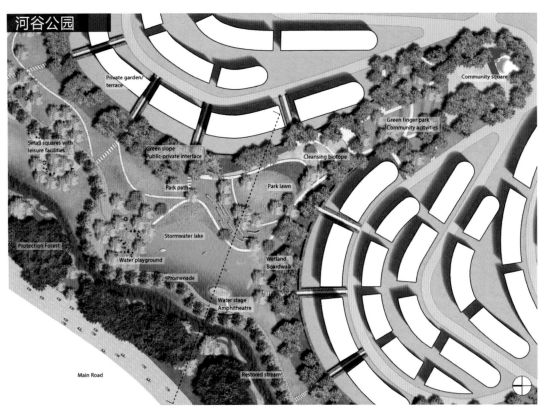

and precipitation in this region result in the drought from May to September. The project harvest rainwater during wet season and allocate these water for urban utilization and irrigation.

Stormwater management system (and sponges urban design) stimulates eco-sustainable development in lattice Omar Pasa district Gaziosmanpasa area, not only manage their water resources and improve the water quality, but also extends to the aspect of water storage for the benefit of the whole city.

Livable City

Urban landscape design and urban design sponge envisioned in providing more services to the urban development, making the city a livable place. It combines more ecological environment design techniques, more biodiversity, more natural water landscape, better water quality and more recreational areas. It allows people to live an extraordinary life in the dense urban environment, but also makes the region become more resilient and flexible when in the face with climate change and natural resources crisis.

A successful livable city construction requires a balance between investment and benefit. A livable environment could improve the value of local real estate obviously. Attracting talents with well-educated background and excellent enterprises, making sure healthy and sustainable growth of the local economy. This is a necessary prerequisite for high quality life in which the people advocated and anticipated.

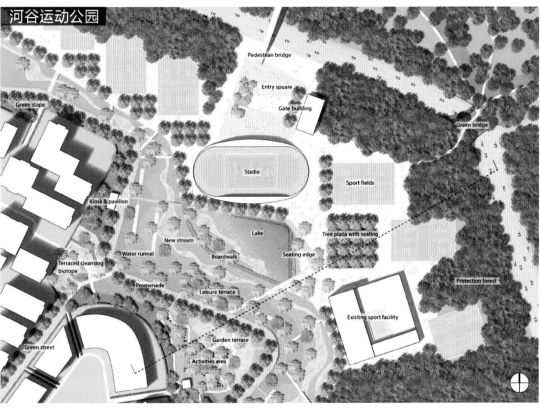

278

加州科学博物馆

CALIFORNIA ACADEMY OF SCIENCES

项目位置：美国加利福尼亚州

项目规模：1 ha

设计公司：SWA集团

获奖信息：2008年10月8日荣获"美国绿色建筑白金奖"

自2008年"绿色建筑节能奖"设立以来，成为世界上体量最大的荣获"公共绿色建筑节能白金奖"的项目

翻　　译：张旭

Location: San Francisco, California, USA

Size: 1 ha

Design Firm: SWA Group

Award: Green Certified LEED Platinum, October 8, 2008

The Largest Public LEED Platinum Energy-saving Building in the World from 2008

Translated by Zhang Xu

项目描述

该项目在设计之初被定位为一栋实体建筑,可容纳现有的多功能构建筑物,包括斯坦哈特水族馆与莫瑞森天文馆,并且在该建筑中碳足迹要有所降低,多余的空间被打造成一个小花园。在设计中,自然景观被架构到三层楼的高度,布满植被的屋顶附着在七个土堆之上。土堆的倾斜夹角虽然足足有60°,但仍与底层景观保持一致。屋顶面积约1 ha,是世界领先的绿色屋顶实验设施之一,主要致力于栖息地质量与连通性的研究。该项目着重倡导能源消耗的降低与浇灌用水的减少,并为技术体系与自然系统的和谐共处提供力证。SWA集团为项目提供了全程景观服务,充分听取了公众意见,并且在该过程中进行了动态思考分析。该项目荣获"公共绿色建筑节能白金奖"。

项目重点

(1)世界体量最大的"双料白金"绿色建筑奖项获奖项目。

(2)本土植被和模纹屋顶,吸引着蝴蝶、小鸟和昆虫。

(3)经过设计的屋顶植被,可不依赖于机械灌溉,在大自然中繁茂生长。

(4)排列在屋顶之上具有光和效应的结构单元可吸收阳光,为博物馆提供能源支持。

Project Description

The new Academy of Sciences building was designed to house multiple previous buildings including the Steinhart Aquarium and Morrison Planetarium in one consolidated building with a smaller overall footprint and with the extra space re-established as gardens for the park. The design lifts the natural landscape three stories up as the dramatic living roof with seven mounds with slopes as steep as 60 degrees conforming to the uses below. The 1-hectare roof is one of the world's leading green roof research facilities, emphasizing habitat quality and connectivity. The project cites reductions in energy consumption and water usage and stands as an example of technical and natural systems coming together in a harmonious way. SWA provided full landscape architectural services for the new Academy of Sciences and the public was carefully listened to and considered during the process. The project has received LEED Platinum certification.

Project Vitals

(1)The largest "Double Platinum" LEED building in the world.

(2) Native plants and green roof, attract butterflies, birds, and insects.

(3) Roof is designed to thrive on natural, not mechanical irrigation sources.

(4) Photovoltaic cells line the roof to collect solar energy that helps power the museum.

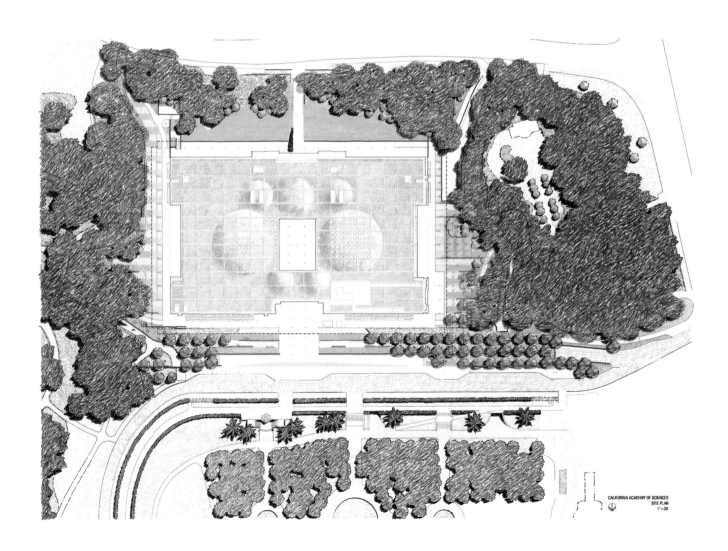

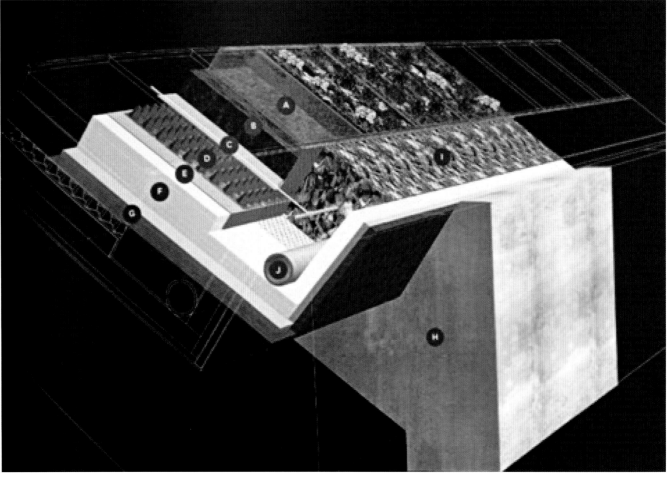

284
伦敦麦克拉伦技术中心水管理项目

WATER MANAGEMENT IN MCLAREN TECHNOLOGY CENTER, LONDON

项目位置：英国伦敦
项目规模：1.62 ha
设计公司：安博·戴水道

Location: London, UK
Size: 1.62 ha
Design Firm: RAMBOLL STUDIO DREISEITL

由于英国的干旱和洪涝现象变得愈发极端和明显,因此该项目采用一种防止环境被破坏且对环境有利的方式,将新的F1赛车调试研发中心设立在自然保护区内成为项目面临的一个大挑战。戴水道设计公司与福斯特建筑事务所合作,创建了一个灵巧、生态的水系统,将雨水管理、生态保护与建筑冷却系统相结合。

　　屋顶和停车场地上的雨水被收集起来并储存于湖中。湖水通过自然生态群落系统在VIP通道下层循环,流至建筑热交换器,并从一个200 m长、极具魅力的水瀑倾泻而下。该自然冷却系统的应用,避免了大型机械冷却塔的建造所造成的环境破坏,保护了场地内的整体生态环境以及景观品质。清澈的湖水为临接河流提供了补给,缓解了溪流的季节性干涸,为地区的生态环境作出了颇有价值的贡献。建筑物和湖体之间的阴阳造型模糊了室内外空间的界限,无论是向建筑内部望去还是从室内远眺,其视觉感观都让人难以忘怀。

As the United Kingdom becomes increasing divided through lines of drought and deluge, it was an essential challenge to site the new Formula 1 research and development centre into a protected natural site in a way that not just limited damage to the environment but actually positively contributes to it. Atelier Dreiseitl collaborated with Foster + Partner and the project team to create a smart, ecological water system which combines rainwater management and ecological restoration with the testing wind tunnels cooling system.

Stormwater run-off from the roof and parking lots is collected and stored in the lake. Lake water is circulated under the VIP access road through a natural planted biotope system, and from there to the building heat-exchanger and back out through a stunning 200 meters long cascade. Thanks to this natural cooling system the need for massive cooling towers was eliminated and thus the landscape character of the site protected. The adjacent stream is fed with clean water from the lake, a valuable ecological contribution in a country where streams are catastrophically running dry. The yin-yang of the building and lake blurs the line between interior and exterior spaces, making views to and from the building unforgettable.

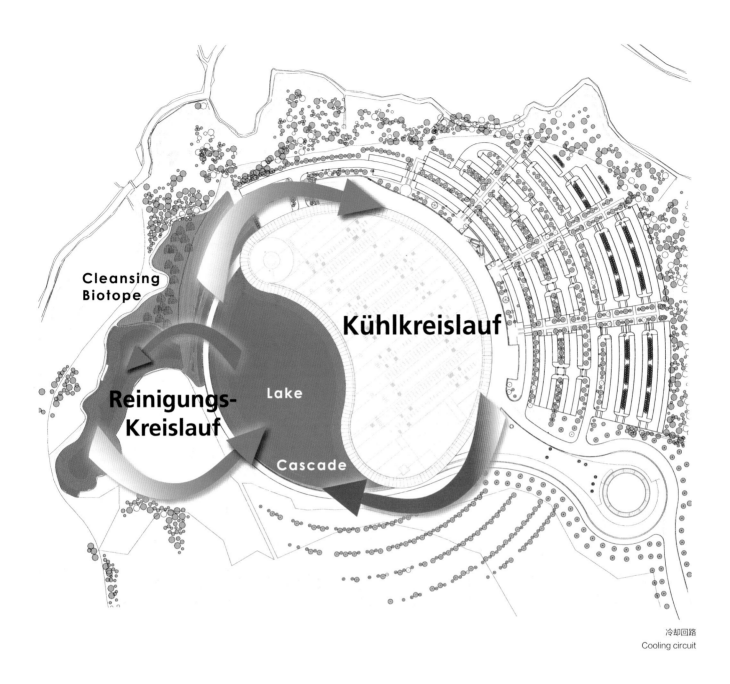

冷却回路
Cooling circuit

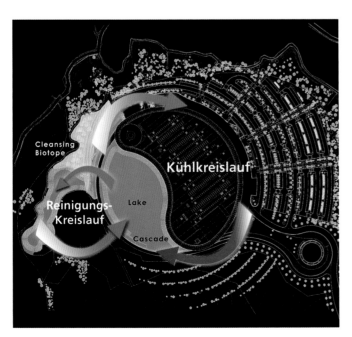
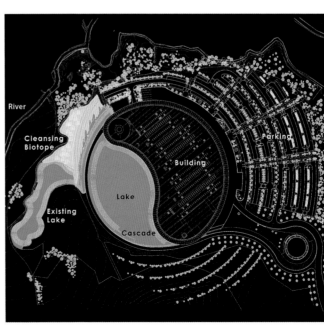

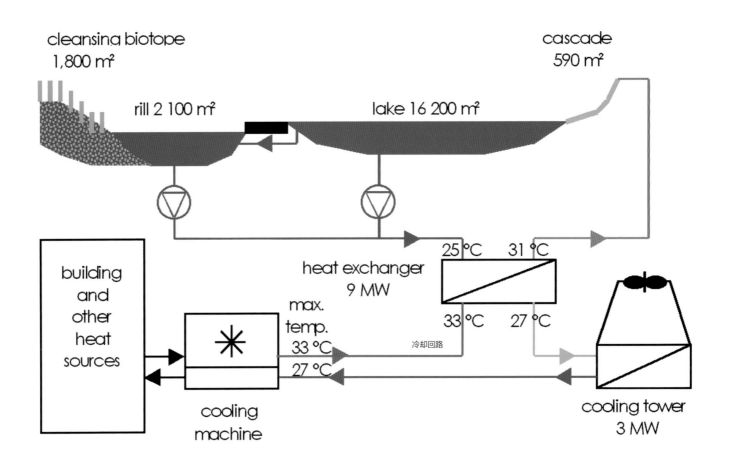

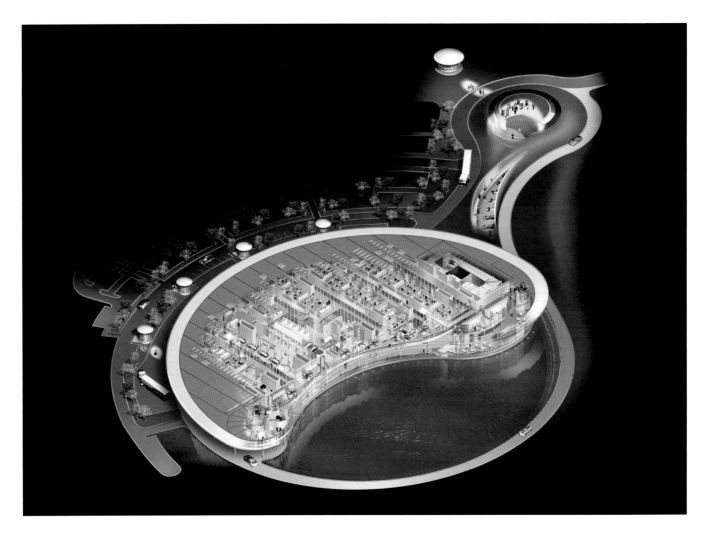

290

中式"方圆"
CHINESE SQUARE & ROUND

项目位置：法国肖蒙市
项目规模：100 m²
设计公司：土人景观，北京大学建筑与景观设计学院

Location: Chateau Chaumont,
FranceSize: 100 m²
Design Firm: Turenscape, Department of Architecture and Landscape, Peking University

该项目是2013法国肖蒙创意园林展的作品,并作为永久作品予以保留。该项目是对传统中式园林的当代解读,整体为外方内圆的形式设计,通过围合空间和小中见大的传统中式园林手法以及填挖方的工程技术,将当代雨水利用理念及传统造园哲学相结合,营造亲切而富有美感的观赏和体验空间。

中传统中式园林的建造基于对大自然的模仿或一种"缩微"的自然景观。这些园林通常由文人和士大夫设计兴建,试图在围合的小空间中再现自然,以人工山水、亭台楼阁、小径曲桥和花木作为元素。空间上,运用"盒子套盒子"的技术和"小中见大"的策略,在有限的空间中,创造无限的体验和风景。与同时代的西方园林的节点和放射视线和路径有截然不同。该项目运用当代设计语汇,重新解读传统中式园林的哲学、手法与体验,同时融入当代生态理念,包括雨洪管理的理念。该设计上有以下几大特点。

(1)挖填方技术和雨洪利用:该场地处于整个园区的低洼地带,通过挖填方,在场地中形成一个方形水池。将土方对应四周,形成围合的高亢之地,便于种植竹林,已形成围合的方形空间。水池反射天光,流动的浮云和更替日月以及四季变幻的植被,在有限的空间中营造了无限的风景。

(2)曲线形的木栈道对角穿越方形水池,漂浮于水面。小栈道仅50 cm宽,容一人穿行。它是"圆"的符号,也是体验水池无限风光的路径。水池中间是三丛红色竹竿,它们取自四周竹林的老杆,是可再生的材料。垂直的竹竿丛沿木栈道分布,令行走在栈道上的游客体验穿越的快感。垂直的竹丛倒映水中,与变换的天景交相辉映,营造了宁静和深远的氛围。水深虽然只有30 cm,但通过垂直向天的红色竹竿,营造了无限深远的感觉。"方圆"周边和水中植物全部为中式植物,渲染出浓郁的中式气氛。

受"太极"图案的启发,方形的池塘通过一条曲折的栈道一分为二。极具中式特色的"太极"图蕴含万事万物皆流转轮回之意,比如年月日、季节、甚至人的生命。方与圆这两种看似相异的两个形状,相得益彰而非冲突对立。方与圆,直与曲,饱和的红色与变换的水色和天空,生机勃勃的绿色竹丛与刺上天空的红色竹竿,从元素到空间,"方圆"让人体验到的是中国:是传统的中国,更是当代的中国!

红色竹竿倒影在水中,营造无限深远的视觉感受。
The reflection of the red bamboo sticks provides a visual experience of infinite extension.

外形取法自"太极图"的木栈道,蜿蜒在方形水池中,象征着万物循环往复,生生不息,既扩大了体验空间,又延展了时间概念。
The meandering walkway is inspired by Tai Chi Symbol, simulating the underbrush and recycling world. It not only amplifies the space, but also expands the concept of time.

水池的方为静,静水渊深;栈道的曲为动,灵动绵长。游客在动静之间,俯仰天地。
The square water is static with depth, while the winding walkway is dynamic. People are able to feel the energy from the contrast between statics and dynamics, and to see the world in this small space.

水池中无限的天光水色,令游人流连忘返。
The reflection of sky and clouds in the water is changing all the time.

自然竹林与人工红竹相映成趣,古典与现代完美融合。
The green of natural bamboo and the red of painted bamboo contradict but coordinate with each other, demonstrating a perfect combination of tradition and modern.

This project was installed in 2013 in Chteau Chaumont, France, and was preserved permanently after the exhibition. This small rain garden, Square & Round, representing earth and sky, is a contemporary re-interpretation of Chinese traditional gardens by applying the formal language of curvilinear and square, the spatial experiential strategy of enclosure and making small into big, as well as the cut-and-fill technique. It integrates the contemporary concept of storm water management with the Chinese gardening philosophy about man and nature, and provides an intimate pleasant environment both for view and experience.

Traditional Chinese garden represents natural landscapes in miniature. Usually built by scholars, poets, and former government officials, the traditional Chinese garden is always enclosed by a wall and has ponds, rocks, trees, flowers, and assorted pavilions that are connected by winding paths. Spatially, Chinese intend to see big from the small, and developed a taste of closed, box-within-box landscape, compared to the westerners' open, point-to-point and radiant view. The project applies the modern design language to reinterpret the traditional Chinese concept of landscape, producing a new look, both psychologically and physically. Meanwhile contemporary ecological methods were integrated in the design, including storm water management with the following features.

(1) Cut-and-fill technique and storm water management: A rain garden situates right in the center of the project, a depression excavated on site that allows rainwater runoff from the surrounding area. Through the cut-and-fill technique, the mount is formed around the pond with bamboo planted on top, which creates a dense screen to block the pond from its surroundings. The changing reflection in the pond, the flowing clouds and the alternation of day and night, all show an infinite changing view within a limited space.

(2) The winding path crossed the pond diagonally and floated above the water. The wooden path is 50 cm in width, allowing only one person to walk through. It's the symbol for "roundness", but also the way to experience the infinite view. Three groups of red bamboos in the middle of the pond are recycled materials and collected on site. Walking between the sky-shooting bamboos provides a pleasant "winding through" experience. the coordination of reflections between the bamboos and the changing sky, created a tranquil atmosphere. Although the water is only 30 cm deep, it deepened the visual effects with the tall bamboos. All plants in this small garden are of Chinese native, which highlighted its Chinese origin.

The creation of a curved floating path visually dividing the water surface into two zones, is inspired by the Chinese Tai chi symbol, which is based on the idea that everything goes in cycles: the years, the months, the seasons, even human life itself. Through the integration of the two seemingly contrasting forms, Square & Round represents the idea of complementation rather than conflict. Square and round, straightness and windiness, bright red and changing reflection of sky in the pond, vibrant green bamboos and sky-shooting red-bamboo fence, and the space and its each elements, all these are demonstrating China: the traditional one and the contemporary one.

方圆 | 园

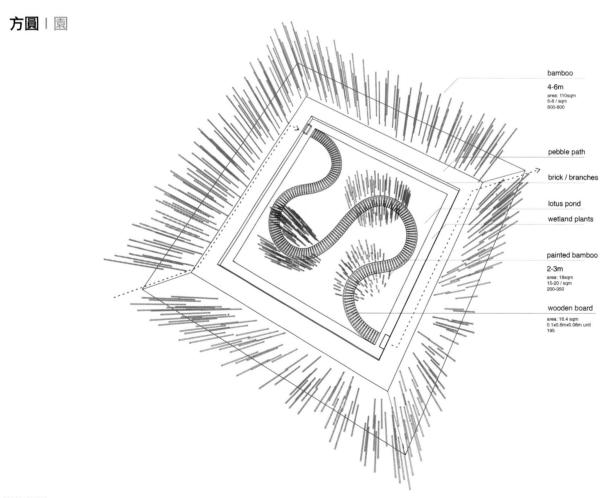

材料细部设计
Detailed material design

方圆 | 园
MATERIAL

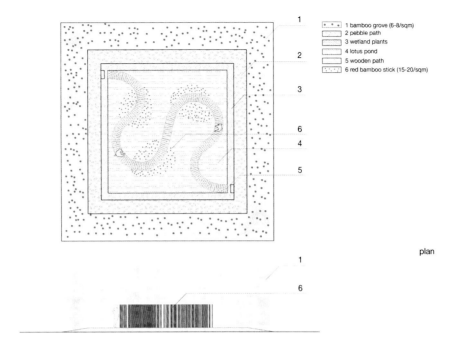

1 bamboo grove (6-8/sqm)
2 pebble path
3 wetland plants
4 lotus pond
5 wooden path
6 red bamboo stick (15-20/sqm)

plan

场地平面图和立面图
Site plan and elevation Plan

方圆 | 园

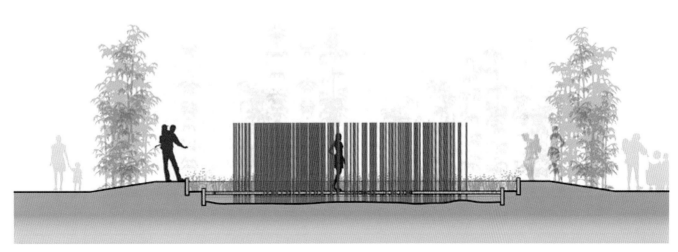

场地剖面图
Section plan

296

清华大学胜因院景观设计

LANDSCAPE DESIGN OF SHENGYIN YUAN, TSINGHUA UNIVERSITY

项目位置：中国北京市清华大学
项目规模：1ha
设计单位：清华大学建筑学院景观学系

Size: 1 ha
Location: Tsinghua University,Beijing,china
Design Firm: Department of Landscape Architecture, Tsinghua University

胜因院是清华大学近现代教师住宅群之一，始建于1946年，曾有多位清华大学知名教授居住于此。60多年来，随着周围地势因建设而抬高，这里遂成为低洼地带，每逢暴雨便导致严重内涝。对这一颇具历史文化价值的地区进行改造，须将历史保护、景观营造与解决雨洪内涝问题结合在一起。近年来，国外最佳管理实践(BMPs)和低影响开发(LID)等雨洪管理措施日益得到推广，它们均强调最大限度地从源头控制径流，就近处理雨水，以达到减少径流、污染物并控制流速的目的。设计团队基于场地及校园历史风貌变迁，提出了胜因院景观设计定位：①校史教育场所和纪念空间；②具有清华特色的人文科研办公区；③"绿色大学"雨洪管理示范场所。

设计方案把降雨、产流、汇流、入渗、排水与竖向、土壤等分析计算并利用相应雨洪管理技术，与场地文脉、空间序列、功能、形式、植物、活动设计相结合。雨水花园作为核心景观要素，满足了蓄渗设计要求，另一方面其形状灵活地顺应周边建筑、道路，结合石笼、条石台阶、置石、旱溪、木平台及植物等元素，使雨水花园不仅作为一种绿色基础设施发挥削减暴雨径流、减缓内涝、处理面源污染等功能性作用，同时也参与表现场地特征、塑造场所精神、体现场地历史、激发场地活力的过程，塑造具有多功能的景观基础设施，也使场地更新改造成为解决积涝问题与创造新景观的契机。这种低成本、低影响、低技术的景观途径，对我国城市解决雨洪内涝问题的策略选择颇具启发意义。

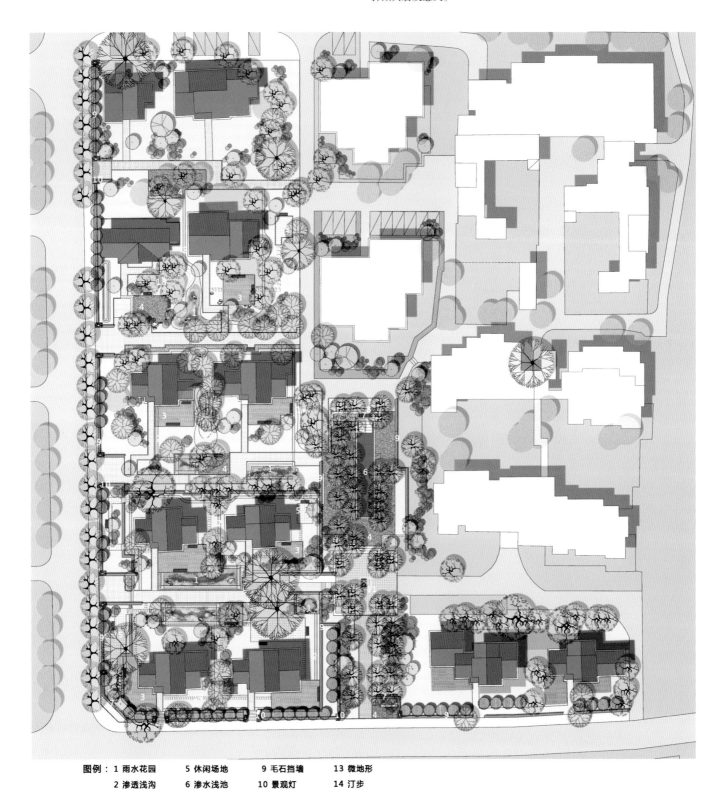

图例：
1 雨水花园　　5 休闲场地　　9 毛石挡墙　　13 微地形
2 渗透浅沟　　6 渗水浅池　　10 景观灯　　14 汀步
3 木平台　　　7 logo景墙　　11 地面浮雕
4 砾石场地　　8 竹丛　　　　12 坡道

雨水花园与毛石矮墙
Rain gardens and rubble parapet

雨水花园台阶边沿与置石
Stairs edge in rain gardens and stone setting

非雨季景观
Dry season landscape

Built in 1964, Shengyin Yuan was one of the modern dwellings for Tsinghua faculties. A number of distinguished professors have lived there, including Mr. Liang Sicheng and Lin Huiyin. But the site has become lower areas with the surrounding construction over 60 years, and often suffers from serious water log problems. In recent years the BMPs, LID and other storm water management techniques have been increasingly accepted and implemented. Based on the history study of the site and campus, the design team puts forward the overall design objective: (1) the campus education and memorial space; (2) the social and humanities science park with university characteristics; (3) the demonstration site of storm water management.

This project illustrates how to solve the on-site storm water issues combined with the representation of the historical context of the site. Firstly, the soil, vegetation, topographic and drainage conditions have been analyzed. And based on the site confluence and infiltration analysis and calculation, several rain gardens are introduced, which has successfully solved the waterlog problem, decreased runoff, encouraged more infiltration, and integrated the local landscape design. Thus the multiple goals including solving the waterlog problem, creating water landscape, and shaping the site spirit have been achieved. Rain gardens and other storm water management facilities not only play the role of green infrastructure, but also become the integral parts of the space to reflect the character, embody the history and arouse the vitality of the site. The landscape approaches of this kind with low cost, low impact and low technique have been proved to be effective for relevant sites to solve the waterlog problems.

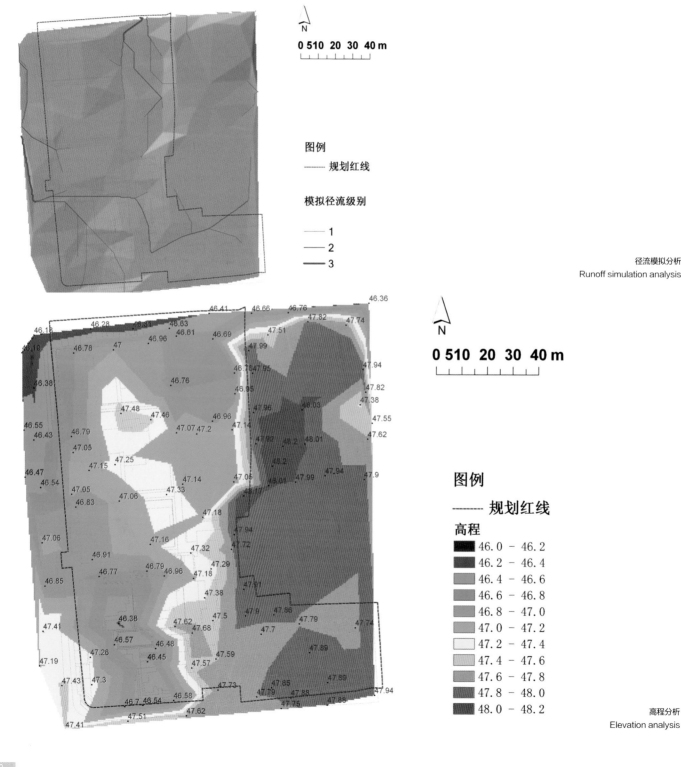

径流模拟分析
Runoff simulation analysis

高程分析
Elevation analysis

积水漫入低洼建筑
Accumulated water diffused into the low-lying buildings

改造前环境杂乱
Cluttered environment before transformation

春季
Spring

夏季
Summer

秋季
Autumn

冬季
Winter

雕塑墙
Sculpture wall

木平台
Wood platform

解说牌
Interpretation board

304

768创意产业园区
阿普雨水花园设计

Design of Up+S Design Rain Garden in the 768 Creative Industrial Park

项目位置：中国北京市
项目规模：170m²
设计公司：阿普贝思（北京）建筑景观设计咨询有限公司
翻　　译：张旭

Location: Beijing, China
Size: 170 ㎡
Design Firm: Up+S Design
Translated by Zhang Xu

阿普雨水花园位于北京林大北路的768创意产业园区内，是阿普贝思国际联合设计机构的户外花园，研究雨水如何排放成为建造阿普雨水花园的最初契机。雨水花园的建设面积170余平方米，通过收集来自办公屋顶140余平方米的屋面雨水以及来自道路上的实时径流，将其汇入花园进行过滤、下渗以及收集。2014年初计划建造，但因用地审批问题，一直延误到2015年4月才建成。

雨水花园是雨洪最佳管理措施（BMPS），也是由雨水基础设施组成的造景系统。它包括花园设施的景观化设计和利用雨水造景的花园。因此，设计师希望改善当前建筑无组织排水状态的同时，减小大雨对场地植物的影响以及市政管网的排水压力。

雨水的组织是一个雨水花园的灵魂。雨落管、开口路牙、台地、景观水池、浅洼绿地以及地下贮水池，共同组成了一个集收集、过滤、下渗、回用于一体的雨水系统。屋面雨水经由雨落管聚集到弃留池，沉淀杂质。后经溢流口优先流入景观水池，景观水池与地下贮水池相连通，通过泵形成动态水景。当雨量较大时，弃留槽内的雨水则通过另一溢流口进入台地，一部分雨水在台地上下渗消解，另一部分形成径流汇入浅洼绿地。设计师对浅洼绿地的垫层进行特殊处理，延缓雨水下渗。如此一来，当浅洼绿地内的雨水水面高于溢流口时，雨水通过管道自动流入地下贮水池。

材料是实现设计理念与雨水组织的关键一环，也是该项目设计中着力探索的一个方面。设计师采用了一些既生态节能又美观耐看的材料。黄色玻璃钢格栅在花园中十分显眼，它在场地原有的洋白蜡及景观水池周边形成一个遮荫的小型停留场地，其质轻、耐久、镂空、易切割等特性不但维持了开发前后现状树周边水热条件的平衡，而且便于其下管线的维护管理。

石笼具有材料易得、施工简易、耐久性强等优点，在场地中进行了两处尝试。一处位于浅洼绿地西侧，作台地之间的挡土。该处使用常规的格宾石笼，填充石材的质地、色彩均与台地其他部分达成统一。另一处即花园入口处的Logo景墙，以蓝色碎玻璃块作为填充，以钢板及钢丝网做支撑。

此外，场地里选择性地种植了大量既耐湿又耐旱的植物，除了保留原有的两棵白蜡和移栽的拂子茅外，还栽植了大量观赏草，如蓝羊茅、崂峪苔草等。

在城市雨洪管理中，景观用地由于建设强度相对较小，为建筑以及硬质场地"基于源头控制"提供了最佳的雨洪消减场所。这也是在小尺度的景观设计中，雨洪设计必须同步考虑、必须为周边雨水服务的原因。雨水花园是海绵城市的细胞体，也是保障城市雨洪安全的基础环节；其应被纳入景观设计体系并作为设计的关键一环。

Up+S Design rain garden is located in the 768 creative industrial park by Beijing Forestry University north road. It is the garden outside the Up+S Design international joint design agency' office. The project try to find out how the issue of rain emissions initiates the construction of the Up+S Design Rain Garden .The rain garden occupies 170 square meters. Stormwater from the 140 square meters office roofs and the timely runoff on roads was conducted and gathered in the garden to be filtered, infiltrated and harvested. The project was planed to build at the beginning of 2014, but for land examination and approval problems, the construction work has been delayed until April 2015.

Rain garden is the stormwater best management practices (BMPS), also a landscape composed of rainwater infrastructure system. It includes landscaping the garden infrastructure and utilizing the stormwater to landscaping. There's no organized discharge system in the building. Therefore, the company wants to improve the current situation, reduce the heavy rains' effects on local plant and relieve the pressure of the municipal pipe drainage network.

The organization of the rain is the soul of the rain garden. The company formed a rainwater system integrated the function of collection, filtration, infiltration and reuse, through the facilities of rain pipe, bifurcated road edge stone, terrace, landscape pond, low-lying green space and underground reservoir. The roof runoff delivered by rain-dropping pipes is gathered in the abandoned conservation pool for the sedimentation and decontamination process. Then the purified rains preferentially run into the landscape pond through the overflow outlet, which is connected by the underground reservoir, and forming a dynamic water landscape by pumps. When the rainfall is large, the rainwater in the abandoned conservation pool flow into the platform from another outlet, a part of the rain infiltrated on the platform ground, the rest gathered into the low-lying green land in the form of runoff. Designers carried on a special treatment to the sub crust layer on which rain infiltration will be delayed. As a result, when the rainwater inside the low-lying green land exceeded the overflow outlet, it would be collected in the underground cisterns automatically by the pipeline.

Material is a key point in the realization of design concept and the organization of stormwater, and also a field the designers want to explore in. Designers want to use ecological and aesthetical materials, but also endurable and energy saving. The yellow grille fences made of glass fiber reinforced plastic are very prominent in the garden. They have created a small shadowing stopover land upon the original Green Ash landscape and the surrounding landscaping pools. This kind of material boasts of the characteristics of lightweight, durable, hollow and easily cutting. All these features not only bestowed the space with the balanced hydrothermal conditions around the trees before and after development, and facilitated its pipeline maintenance and management.

The materials used in stone cage are accessible, the construction work of stone cage is easy, and the feature of stone cage is endurable. Because of all the merits the form of stone cage embodied, two cases were implemented on site. One is located in the west side of the low-lying land, as the earth retaining wall among platforms. The regular gabion stone cages were employed on this site. The filling materials' texture and color of the stone cages match the rest part of the platform brilliantly. Another case is the Logo Wall at the entrance of the garden. The filling materials in stone cages are blue fragmental glass. The supporting structure is made of steel plate and wire mesh.

A large number of wet-resistant and dry-resistant plants are selectively planted on site. In addition to the original remained two Green Ash trees and the transplanted Calamagrostis epigejos, a lot of ornamental grasses were used in this project, namely, Blue Fescue, Carex Giraldiana, and other ten kinds of ornamental flowers and plants.

In urban stormwater management, the construction intensity of the landscape land is relatively small. Therefore, based on the source controlling principle, the site is the best stormwater reduction place for the buildings and hardened place. This is also the reason why the plan must synchronously consider and serve for the stormwater management of the surrounding places in a small-scaled landscape design. Rain garden is the cell of the Sponge City. It is a crucial link in the urban stormwater management security, and shall be incorporated into the landscape design system as a key point.

新加坡 ABC 导则

在营造波光粼粼如画般的河流和湖泊以及优美的河岸这一愿景的驱动下，新加坡接受了转变为"花园与水之城市"的挑战。

根据新加坡国家水务机构公共事业局推出的"活跃、美观、洁净"（ABC）之水域方案，新加坡正在努力发展成为"花园与水之城市"。

结果表明，这项方案已经取得了越来越多的发展成果。新加坡制订的国家愿景，即拥有波光闪闪的河流、美景如画的河岸、溪流中皮船悠闲划过以及洁净的水流汇入如画般的湖泊，正在逐步变为现实。

1. 转变成为花园与水之城市

多年来，新加坡建立的排水管网渗透面非常广泛，包括32条主河流、8000 km长水渠以及17座水库。为了充分挖掘该项水利基础设施的潜力，新加坡国家水务机构公共事业局在2006年推出了"活跃、美观、洁净"（ABC）之水域方案。这项战略举措旨在引导充分挖掘水体潜力，提高水资源的质量和人们的生活水平。ABC水域方案通过将排水沟、水渠和水库与周围的环境相结合营造美观、洁净的溪流、河流和湖泊，并为人们打造如画般的社区活动空间。

2. 保证供水持续、稳定并在人们负担能力范围之内

新加坡一直注重投资研发和技术，通过"国家四大水喉"（即天然降水、进口水、新生水和淡化海水）实现了供水的多样性和稳定性。鉴于该项多目的样性，新加坡关闭了水回路，采取措施提高了水资源的稳定性。同时，这也成为推出ABC水域方案的背景，因为新加坡不仅仅满足于水资源的充分性，还大力投资充分挖掘水资源的潜力，提高人们的生活水平。

3. ABC水域方案

ABC水域方案旨在无缝综合环境（绿色）、水体（蓝色）及社区（橙色），创造新型的社区空间，鼓励人们在水域及其周围地区开展娱乐休闲活动。由于社区更靠近水域，人们将更加珍惜宝贵的水资源，树立保护水资源的意识。

ABC水域方案的三项关键策略包括以下内容：

（1）制订ABC水域主计划和实施项目。

2007年推出的主计划旨在引导整体项目实施情况，将城市的实用下水道、水渠和水库转变为与环境融为一体的活跃、美观、洁净之溪流、河流和湖泊。目前已确定到2030年超过100个预期项目将在岛上分阶段实施完成。自2014年6月，已经实施了23个ABC水域项目。

（2）采纳ABC水域理念。

ABC水域理念诠释了新加坡充分挖掘水域潜力并与环境和生活方式相结合的规划理念。公共事业局开始意识到这项理念的效益，因为许多公共机构和私营企业逐渐在其发展过程中采纳了ABC水域设计景观。这些设计景观正如自然系统，在地面雨水径流汇入水域和水库之前对其进行滞留和净化，同时，还丰富了生物多样性，改善了人们的居住环境。

《ABC水域设计指南》于2009年推出，旨在鼓励公共机构与私营企业加强合作，以适当的方式营造ABC水域设计景观，并在这个过程中改善水域环境，提高水质。

（3）3P（人员部门、公营部门、私营部门）合作方法

如果得不到社区的认同和支持，就无法实现可持续的雨洪管理。公共事业局注重在社区中培养人们的水体主人翁意识。例如，鼓励学校为各项ABC水域项目制订教育学习路径，方便学生进一步了解和珍惜水域。私营企业、基层组织和社区团体同时也帮助改善路径，在ABC水域场所开展多种活动，鼓励更多的人以负责任的态度使用场所及相关设施。

4. 采纳ABC水域理念的优势

ABC水域设计景观通过自然净化过程保护水库和水渠的水质，提高风景的美学性和多样性，减少雨水径流的流动，并使社区中的人们更接近水域，为人们营造新的休闲娱乐空间。

2010年，公共事业局发起了一项新的计划，即ABC水域认证计划，旨在认可ABC水域理念并使ABC水域设计景观被越来越多公共机构和私营企业所采纳。

新加坡建设局（BCA）等若干个政府机构也认可了ABC水域设计景观。作为一项基础计划，BCA绿色标识计划在环境设计和绩效方面融入了国际认可的最佳惯例。该计划包括ABC水域设计景观，其作为一种雨洪管理的最佳方式。

2011年，公共事业局和新加坡工程师协会（IES）推出了ABC水域专业人士项目，力求在ABC水域设计景观方面发挥行业专业人士的专长。该项目的参与者在完成四个核心模块和两个选修模块的测试并符合必要标准之后，将有资格登记成为ABC水域专业人士。该项目得到了新加坡建筑师学会（SIA）和新加坡风景园林师协会（SILA）的大力支持。

（翻译：张旭）

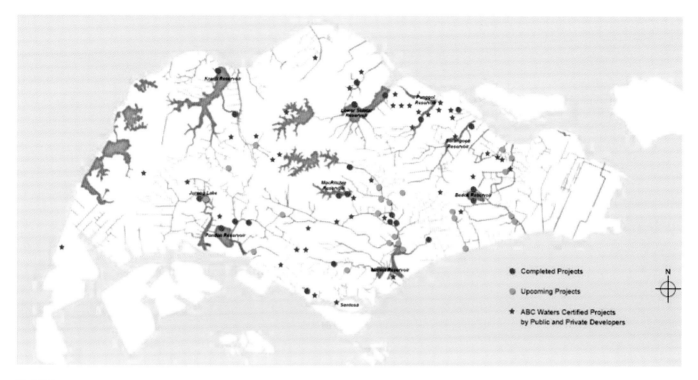

新加坡蓝图
Singapore blueprint

SINGAPORE ABC GUIDELINE

Imaging to be blessed with sparkling river, picturesque lakes and beautiful banks, this vision challenged and motivated Singapore to transformed into a garden city with pleasant scenery.

In line with the Active, Beautiful, Clean (ABC) Water Zone Program proposed by Singapore's National Water Agency & Public Utilities Board (PUB), Singapore is strived to becoming a city of garden and a city of water.

More and more fruits of the city development verified the success achieved in the project planning. In Singapore's vision, the nation could embrace the rivers with weaves and ripples, the lake with lovely banksides. The kayak rowing on the streams slowly and leisurely, clean water continuously flowing into the lakes just like a fairytale. All of this would be accomplished in the near future.

1. Transformed into A City of Garden & Water

Over the years, the drainage pipe system Singapore had installed covers a large amount of areas and sites. It includes 32 main rivers, one 8000-km water channel and 17 reservoirs. In order to fully tap the potential of the hydraulic infrastructure, Singapore national water agency & Public Utilities Board launched the Active, Beautiful, Clean (ABC) Water Zone Program in 2006. This strategy aims at to bring the water potential into full play, improve the quality of the water resource and the standard of citizen's life. In ABC Water Zone Program, the drainage ditches, water channels and reservoirs are integrate in their surroundings, bringing back beautiful clean streams, rivers and lakes, and creating picturesque community activity space for people.

2. Ensure Continuous and Stable Water Supply within Citizen's Affordability

Singapore always attaches great emphasis on R & D (research and development) and T & T (technology and technique) investment. Through the Four Great National Water Throat (natural precipitation, imported water, NEWater and desalinated sea water), Singapore realized the diversity stability of the water supply. Due to the diversified supply resources, Singapore closed the water-loop. Take measures to improve the stability of water resources. Spontaneously this condition became the background in where the ABC Water Zone Program first been proposed. Because Singapore won't just content on the adequacy of water resources, on a further stage, they would proliferate their investment on taking fully advantage of the water resources and improve the people's living standards.

3. ABC Waters Program

ABC Water Zone Program is project to integrate the environmental (green), water (blue) and Community (orange) in a seamless way, create a new type of community space, encourage people to carry out recreational and leisure activity in the waterfront area. Since the communities are closer to the water, people would perceive the water resources in a more cherished and precious way. And People would be more aware of the urgency in protecting the water resource.

The three key strategies of the ABC Waters Program are as follows.

(1) Formulating the ABC Waters master plan and implement the projects.

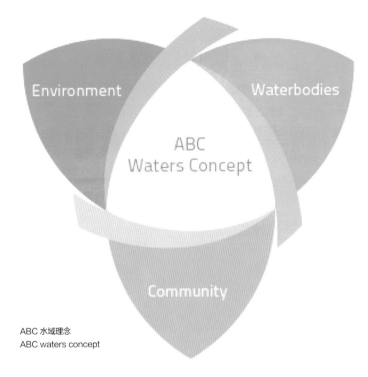

ABC 水域理念
ABC waters concept

The master plan was proposed in 2007, aiming at steering the implement and practice of the overall program. The existing sewer utilities, manmade canals and reservoirs in use were transformed into dynamic beautiful and clean streams, rivers and lakes, which are integrated in the environments. The program anticipates more than 100 identified projects would be completed on the inlands in steps and phrases. By June 2014, 23 projects in the ABC Water Zone Program have been constructed and accomplished.

(2) Advocating in adopting the ABC Water Zone concept.

ABC waters concept interprets the joint idea of fully tapping the potential of waters and combined them with environment protection and lifestyle improvement. Public Utilities Board began to realize the benefits obtained from the concept, because many public institutions and private agencies gradually adopt the idea of the ABC Water Zone landscape design in their development process. These designed landscape is similar to natural system, storm water runoff would be retained and purified before discharging into waters and reservoirs. Meanwhile, the landscape enhanced the ecological diversity and improved the living environment.

ABC Waters Design Guide was published in 2009. It calls for cooperation, and encourage the public-supported business and private-operated enterprises to explore the appropriate ABC waters Landscape design approaches fitting themselves, improve the water environment during their companies growth process.

(3) Encouraging 3P (personnel department, public sector, private sector),cooperation.

The sustainable storm water management cannot be achieved if the community is dissenting and unsupportive. PUB would cultivate their awareness of self-governance and self-management in a consistent way. For instance, they encourage schools to develop and provide the educational access for citizens in the theme of ABC Waters projects, enable and facilitate the students to learn more about the program, and cherish waters in a more conscious way. Meanwhile, private companies, grassroots organizations and community groups also helped to improve and enlarge the ABC waters project's learning path, and carry out a variety of activities on ABC program sites, encourage people to behave in an more responsible manner when they are using the sites and related facilities.

4. The Superiority in Adopting Concept ABC Waters

To protect the water quality in reservoirs and canals through natural purification process, the ABC Waters landscape design has leapt forward in improving the aesthetics and diversity on the site, reducing the flow volume of the stormwater runoff, as well as makting the community more intimate to the water zones, and creating new recreational spaces for people.

Public Utilities Board has launched a new program in 2010, which was the ABC Water Certification Scheme. This newly-planed program was dedicated in approving and adopting the ideas of ABC waters, and propagating the concept of ABC Waters landscape design to public institutions and private developers who sponsored in development activities.

The ABC Waters landscape design has gained the recognitions from the Building & Construction Authority of Singapore (BCA) and other official agencies. As a basic plan, BCA Green Labeling Programs have employed the best practice of which are internationally recognized in environment design and performance operation. The ABC Waters landscape design is included in the plan as one of the best stormwater management measures.

The Public Utilities Board and Institute of Engineers Singapore (IES) launched the ABC Waters Professionals Project in 2011, which is devoting to cultivate the expertise and professionals in the discipline of ABC Waters landscape design. Participants would be eligible to register as an ABC Waters professional after they finished four compulsory and two elective modular curriculums and meet the necessary criteria of the courses. Besides, the project was recognized and supported by the Singapore Institute of Architects (SIA) and the Singapore Institute of Landscape Architects (SILA).